AI 速成课程(影印版)
AI Crash Course

Hadelin de Ponteves 著

南京　东南大学出版社

图书在版编目(CIP)数据

AI 速成课程:影印版:英文/(法)哈德琳·德.庞特
维斯著. —南京:东南大学出版社,2020.8
书名原文:AI Crash Course
ISBN 978-7-5641-8970-9

Ⅰ.①A… Ⅱ.①哈… Ⅲ.①人工智能-英文
Ⅳ.①TP18

中国版本图书馆 CIP 数据核字(2020)第 116741 号
图字:10-2020-181 号

AI 速成课程(影印版)

出版发行:东南大学出版社
地　　址:南京四牌楼 2 号　　邮编:210096
出 版 人:江建中
网　　址:http://www.seupress.com
电子邮件:press@seupress.com
印　　刷:常州市武进第三印刷有限公司
开　　本:787 毫米×980 毫米　　16 开本
印　　张:22.5
字　　数:441 千字
版　　次:2020 年 8 月第 1 版
印　　次:2020 年 8 月第 1 次印刷
书　　号:ISBN 978-7-5641-8970-9
定　　价:108.00 元

本社图书若有印装质量问题,请直接与营销部联系。电话(传真):025-83791830

packt.com

Subscribe to our online digital library for full access to over 7,000 books and videos, as well as industry leading tools to help you plan your personal development and advance your career. For more information, please visit our website.

Why subscribe?

- Spend less time learning and more time coding with practical eBooks and Videos from over 4,000 industry professionals

- Learn better with Skill Plans built especially for you

- Get a free eBook or video every month

- Fully searchable for easy access to vital information

- Copy and paste, print, and bookmark content

Did you know that Packt offers eBook versions of every book published, with PDF and ePub files available? You can upgrade to the eBook version at www.Packt.com and as a print book customer, you are entitled to a discount on the eBook copy. Get in touch with us at customercare@packtpub.com for more details.

At www.Packt.com, you can also read a collection of free technical articles, sign up for a range of free newsletters, and receive exclusive discounts and offers on Packt books and eBooks.

Contributors

About the author

Hadelin de Ponteves is the co-founder and CEO at BlueLife AI, which leverages the power of cutting-edge Artificial Intelligence to empower businesses to make massive profits by optimizing processes, maximizing efficiency, and increasing profitability. Hadelin is also an online entrepreneur who has created 50+ top-rated educational e-courses on topics such as machine learning, deep learning, artificial intelligence, and blockchain, which have reached over 700,000 subscribers in 204 countries.

First and foremost, I would like to thank Jan Warchocki for his amazing contribution to the book. Jan is one of my top talented students on Udemy, and I had the pleasure to work with him on the creation of this book. He provided very smart feedback on every chapter, added many brilliant examples to illustrate the theory and practice for each of the AI models, and even brought his own practical application in the last chapter, covering the most advanced AI model of this book. Thanks to Jan, the book is richer in relevant AI applications, and more complete in the theory. I give him my highest thanks, and I have no doubt he will become a very talented AI scientist with whom I will have the pleasure to collaborate again on other projects.

Then I would like to thank my business partner and dear friend Kirill Eremenko, who encouraged me to write this book. First, Kirill gave me the inspiration to write a complete book on AI. But most importantly, this book would never have been written without him. Indeed, I was able to create this book thanks to the many years of collaboration I had with Kirill producing our online courses.

This book includes the top explanations and practical activities of our courses, while seriously taking into account all of the most recurrent feedback we collected from our students over the years.

I would also like to thank Alex Patterson, Jonathan Malysiak, and David Barnes from the team at Packt. Alex did a tremendous job improving the book, by reviewing and providing feedback on every single paragraph, making sure everything is top quality from start to finish. Jonathan was like a coach to me, helping me all the way to make this book better and better. Sometimes some big challenges had to be taken on during the creation of this book, and Jon was here to encourage me every time to succeed in them. I am very grateful to him for pushing me to write this book and achieve my highest potential. David was the one to offer me this opportunity in the first place. He came to me with a great idea, convincing me that it was possible to convey the energy I have in my online courses into a book. I accepted the challenge, and today I can say that he was right. Thanks, David, for this brilliant idea.

And finally, I would like to thank the technical reviewers of this book, who not only provided feedback on every chapter of the book, but also made sure that all the code ran correctly. Thanks to them the reader is better guided in the theory of the AI models, and also on how to run all the different code in this book.

About the reviewers

Valeriy Babushkin is the director of Data Science at X5 Retail Group, where he leads a team of 80+ people in the area of natural language processing, machine learning, computer vision, data analysis, and A/B testing. Valeriy is a Kaggle competition grand master, ranking globally in the top 30, and also leads a data science team at Yandex. He studied cybernetics in Moscow Polytechnical University and mechatronics at Karlsruhe University of Applied Sciences and worked with Packt as an author of the *Python Machine Learning Tips, Tricks, and Techniques* course.

Shrinivas Shetty is a budding data science and machine learning enthusiast. He is a Master of Science graduate from Stevens Institute of Technology with a focus in data engineering and business intelligence and analytics. He has been working in the field of AI for a year, and he helps businesses solve problems and improve efficiency using AI and machine learning techniques. He provides software services using programming tools like Python and Scala to develop data science projects and deploy them using data engineering principles. Apart from doing computer science and using AI and machine learning to solve business use cases, he indulges his long-time passion for e-sports (DOTA 2 and CS: GO anyone?).

I would like to thank my family (mom, dad, and younger brother) for being a constant source of support and encouragement. I would also like to thank my dear friends for inspiring me to push through my limitations. It is a privilege to be in your company and a fortune to have your support.

Table of Contents

Preface

Hello, data scientists and AI enthusiasts. For many years I've created online courses on Artificial Intelligence (AI), which have been very successful and contributed well to the AI community. However, something essential was missing. At one point, so many AI courses were made that most of my students asked me for guidance on how to take the courses. So instead of providing an order in which to take the courses, I decided to create an all-in-one full guide to AI as a book, which would include in a perfect structure all the best explanations and real-world practical activities from my courses.

You see, my goal is to democratize AI and raise awareness among everyone of the fact that AI is an accessible technology that can make a difference for the better in this world. I am trying my best to spread knowledge around the world to get people prepared for the future jobs and opportunities of this 21st century. And I thought some people would learn AI much more efficiently from an all-in-one book they can take anywhere, rather than completing tens of online courses that can be hard to navigate. That being said, this book is also a great additional resource for those people who do prefer, and take, online courses.

My simple hope for this book is that more people learn AI the right way, as a result of me offering them this efficient alternative to online courses. I've succeeded at the challenge of including the best of my training in a single book, and today I'm truly happy to release it. I sincerely hope it will help more people land their dream job, grow an amazing career in data science or AI, and bring beautiful solutions to the tough challenges of this 21st century.

Who this book is for

Anyone interested in machine learning, deep learning, or AI.

People who aren't that comfortable with coding, but who are interested in AI and want to apply it easily to real-world problems.

College or university students who want to start a career in data science or AI.

Data analysts who want to level up in AI.

Anyone who isn't satisfied with their job and wants to take the first steps toward a career in data science.

Business owners who want to add value to their business by using powerful AI tools.

Entrepreneurs who are eager to learn how to leverage AI to optimize their business, maximize profitability, and increase efficiency.

AI practitioners who want to know what projects they can offer to their employees.

Aspiring data scientists, looking for business cases to add to their portfolio.

Technology enthusiasts interested in leveraging machine learning and AI to solve business problems.

Consultants who want to transition companies into being AI-driven businesses.

Students with at least high school knowledge in math, who want to start learning AI.

What this book covers

Chapter 1, Welcome to the Robot World, introduces you to the world of Artificial Intelligence.

Chapter 2, Discover Your AI Toolkit, uncovers an easy-to-use toolkit of all the AI models as Python files, ready to run thanks to the amazing Google Colaboratory platform.

Chapter 3, Python Fundamentals – Learn How to Code in Python, provides the right Python fundamentals and teaches you how to code in Python.

Chapter 4, AI Foundation Techniques, introduces you to reinforcement learning and its five fundamental principles.

Chapter 5, Your First AI Model – Beware the Bandits!, teaches the theory of the multi-armed bandit problem and how to solve it in the best way with the Thompson Sampling AI model.

Chapter 6, AI for Sales and Advertising – Sell like the Wolf of AI Street, applies the Thompson Sampling AI model of *Chapter 5* to solve a real-world business problem related to sales and advertising.

Chapter 7, Welcome to Q-Learning, introduces the theory of the Q-learning AI model.

Chapter 8, AI for Logistics – Robots in a Warehouse, applies the Q-learning AI model of *Chapter 7* to solve a real-world business problem related to logistics optimization.

Chapter 9, Going Pro with Artificial Brains – Deep Q-Learning, introduces the fundamentals of deep learning and the theory of the deep Q-learning AI model.

Chapter 10, AI for Autonomous Vehicles – Build a Self-Driving Car, applies the deep Q-learning AI model of *Chapter 9* to build a virtual self-driving car.

Chapter 11, AI for Business – Minimize Cost with Deep Q-Learning, applies the deep Q-learning AI model of *Chapter 9* to solve a real-world business problem related to cost optimization.

Chapter 12, Deep Convolutional Q-Learning, introduces the fundamentals of convolutional neural networks and the theory of the deep convolutional Q-learning AI model.

Chapter 13, AI for Games – Become the Master at Snake, applies the deep convolutional Q-learning AI model of *Chapter 12* to beat the famous *Snake* video game

Chapter 14, Recap and Conclusion, concludes the book with a recap of how to create an AI framework and some final words from the author about your future in the world of AI.

To get the most out of this book

- You don't need to know much before we begin; the book contains refreshers on all the prerequisites needed to understand the AI models. There's also a full chapter on Python fundamentals to help you learn, if you need to, how to code in Python.

- There are no required prior installations, since all the practical instructions are provided from scratch in the book. You only need to have your computer ready and switched on.

- I recommend you have Google open while reading the book, so that you can visit the links provided in the book as resources, and to check out the math concepts behind the AI models of this book in more detail.

Download the example code files

You can download the example code files for this book from your account at `http://www.packtpub.com`. If you purchased this book elsewhere, you can visit `http://www.packtpub.com/support` and register to have the files emailed directly to you.

You can download the code files by following these steps:

1. Log in or register at `http://www.packtpub.com`.
2. Select the **SUPPORT** tab.
3. Click on **Code Downloads & Errata**.
4. Enter the name of the book in the **Search** box and follow the on-screen instructions.

Once the file is downloaded, please make sure that you unzip or extract the folder using the latest version of:

- WinRAR / 7-Zip for Windows
- Zipeg / iZip / UnRarX for Mac
- 7-Zip / PeaZip for Linux

The code bundle for the book is also hosted on GitHub at `https://github.com/PacktPublishing/AI-Crash-Course`. We also have other code bundles from our rich catalog of books and videos available at `https://github.com/PacktPublishing/`. Check them out!

Download the color images

We also provide a PDF file that has color images of the screenshots/diagrams used in this book. You can download it here: `https://static.packt-cdn.com/downloads/9781838645359_ColorImages.pdf`.

Conventions used

There are a number of text conventions used throughout this book.

`CodeInText`: Indicates code words in text, database table names, folder names, filenames, file extensions, pathnames, dummy URLs, user input, and Twitter handles. For example; "To get these numbers you can add together the lists `nPosReward` and `nNegReward`."

A block of code is set as follows:

```
# Creating the dataset
X = np.zeros((N, d))
for i in range(N):
    for j in range(d):
        if np.random.rand() < conversionRates[j]:
            X[i][j] = 1
```

When we wish to draw your attention to a particular line in a code block, we have included the line numbers so that we can refer to them with precision:

```
80          self.last_state = new_state
81          self.last_action = new_action
82          self.last_reward = new_reward
83          return new_action
```

Any command-line input or output is written as follows:

```
conda install -c conda-forge keras
```

Bold: Indicates a new term, an important word, or words that you see on the screen, for example, in menus or dialog boxes, also appear in the text like this. For example: "Select **System info** from the **Administration** panel."

Warnings or important notes appear like this.

Tips and tricks appear like this.

Get in touch

Feedback from our readers is always welcome.

General feedback: Email feedback@packtpub.com, and mention the book's title in the subject of your message. If you have questions about any aspect of this book, please email us at questions@packtpub.com.

Errata: Although we have taken every care to ensure the accuracy of our content, mistakes do happen. If you have found a mistake in this book we would be grateful if you would report this to us. Please visit, http://www.packtpub.com/submit-errata, selecting your book, clicking on the Errata Submission Form link, and entering the details.

Piracy: If you come across any illegal copies of our works in any form on the Internet, we would be grateful if you would provide us with the location address or website name. Please contact us at copyright@packtpub.com with a link to the material.

If you are interested in becoming an author: If there is a topic that you have expertise in and you are interested in either writing or contributing to a book, please visit http://authors.packtpub.com.

Reviews

Please leave a review. Once you have read and used this book, why not leave a review on the site that you purchased it from? Potential readers can then see and use your unbiased opinion to make purchase decisions, we at Packt can understand what you think about our products, and our authors can see your feedback on their book. Thank you!

For more information about Packt, please visit packtpub.com.

Welcome to the Robot World

1

"We are truly living in the most exciting time to be alive!" These words, by the great tech entrepreneur Peter Diamandis, are even more true for people working in the **artificial intelligence** (**AI**) ecosystem. There is a reason why AI jobs are considered the sexiest jobs of the 21st century: besides being very well paid, AI is a fantastic topic to work on.

AI is taking a more and more important place in the world, and today we can find applications of it in almost all industries. This is not a temporary trend; AI is here to stay. As the top AI leader and influencer Andrew Ng said, AI is the new electricity. Just like the industrial revolution transformed lives and jobs in the 19th century, AI is about to do the same in this 21st century. Hence, the more you understand and know how to use it, the more opportunities will open up to you.

To give you some important figures, according to a study done by **PricewaterhouseCoopers** (**PwC**), AI could contribute up to $15.7 trillion to the global economy by 2030, which is more than the current output of China and India combined. So, you've definitely made a great choice to study this field. Welcome to the incredible world of Artificial Intelligence!

In this chapter, you will begin your AI journey with a top-level view of everything you'll learn from this book as you read and work through the chapters ahead with me. Then, I'll help you understand where learning AI can take you, by going through a variety of top industry applications for Artificial Intelligence.

Beginning the AI journey

Being a young AI scientist, I remember my first days in AI very well. This is important because this book is a crash course in AI. You don't need any prior knowledge of the field to work through the chapters.

In this book, I will explain the solid foundations of AI, while making sure to answer all the questions that I had back when I started in this field in detail. This means that everything will be explained step by step, and your learning process will follow a smooth path, supported by the relevant logic.

Having the right information at your fingertips is not enough to successfully break into the AI world. What you also need is energy, enthusiasm, and excitement. Even better, you need passion, and ideally obsession, about the subject. As an experienced tutor of online courses, I hope to pass on my knowledge and, most importantly, my passion.

In this book, you will go on a journey together with me, taking a path through a world of exciting AI applications, including many real-world case studies in the chapters. The applications will follow an increasing level of difficulty, from the simplest model in AI, to a much more advanced level.

For each of the AI applications, I will focus mostly on the intuition needed to understand them, and then, for those interested in the mathematics and pure theory behind the application, I will provide those as an option. The reason why I choose to focus on intuition rather than math is not only because I want to make this book easy to understand for everyone, but also because, in order to perform well in AI today, it is extremely important to have the right intuition. When you're solving a problem with AI, you have to figure out which model best fits your problem environment, and you can only do that when you have the proper intuition of how each AI model works.

Four different AI models

These AI models were chosen to be part of this book because they are used in a great variety of industry applications and can solve many different real-world problems. I'll just reveal their names here before we study them in depth across the book. The four AI models you will learn everything about in this book are the following:

1. Thompson Sampling
2. Q-learning
3. Deep Q-learning
4. Deep convolutional Q-learning

For each of these four models, we will follow the same three-step approach:

1. Get an intuitive understanding of how it works.
2. Get all the math behind the theory.
3. Implement the model from scratch in Python.

I have followed this structure many times with my students, and I can tell you that it works the best. The idea is simple: because you start with your intuition, you won't get overwhelmed by the math, but will instead understand it more easily. You'll also feel comfortable coding some models of which you both have an intuitive understanding and in-depth theoretical knowledge.

The models in practice

All the way through this book you'll find practical examples to learn from or implement yourself. Here's a list of the AI implementations you'll find in the chapters of this course, which start in *Chapter 3* after you get the tools you need for your AI journey in *Chapter 2*.

Fundamentals

Chapter 3, Python Fundamentals – Learn How to Code in Python, contains the Python coding fundamentals you'll need for this book. You can remind yourself, or learn from scratch, how to code in Python.

Chapter 4, AI Foundation Techniques, contains a pseudocode example to illustrate the five core principles of Artificial Intelligence.

Thompson Sampling

Chapter 5, Your First AI Model – Beware the Bandits!, contains introductory code to illustrate the theory behind the Thompson Sampling AI model.

Chapter 6, AI for Sales and Advertising – Sell like the Wolf of AI Street, contains a real-world implementation of the Thompson Sampling model, applied to online advertising.

Q-learning

Chapter 7, Welcome to Q-Learning, contains pseudocode to illustrate the theory of the Q-learning AI model.

Chapter 8, AI for Logistics – Robots in a Warehouse, contains a real-world implementation of the Q-learning model, applied to process automation and optimization.

Deep Q-learning

Chapter 9, Going Pro with Artificial Brains – Deep Q-Learning, contains introductory code to illustrate the theory behind Artificial Neural Networks.

Chapter 10, AI for Autonomous Vehicles – Build a Self-Driving Car, contains a real-world implementation of the deep Q-learning model, applied to self-driving cars.

Chapter 11, AI for Business – Minimize Costs with Deep Q-Learning, contains another real-world implementation of the deep Q-learning model, applied to energy and business.

Deep convolutional Q-learning

Chapter 12, Deep Convolution Q-Learning, contains introductory code to illustrate the implementation of a **Convolutional Neural Network (CNN)**.

Chapter 13, AI for Video Games – Become the Master at Snake, contains a real-world implementation of the deep convolutional Q-learning model applied to a game.

As you can see, every time you're introduced to a new model, you learn the intuition first, then the math, and then you move to an implementation of the model. So, why is learning how to implement these models worth your while?

Where can learning AI take you?

I'd like to motivate you by showing you that you made the right choice to learn AI. To do this, I'll take you on a tour of all the incredible applications AI can and will have in the 21st century. I have a vision of how AI can transform the world, and this vision is structured around 10 areas.

Energy

In 2016, Google used AI to reduce energy consumption in its data centers by more than 30%. If Google has done it for data centers, it could be done for an entire city. By building a smart AI platform using **Internet of Things (IoT)** technology, the consumption and distribution of energy can be optimized on a large scale.

Healthcare

AI has enormous promise for healthcare. It can already diagnose diseases, make prescriptions, and design new drug formulas. Combining all these skills into a smart healthcare platform will allow people to benefit from truly personalized medical care. This would be amazing for society. The challenges in achieving this are not only present in the technology, but also in getting access to anonymous patient data, which so far is protected by regulations.

Transport and logistics

Self-driving vehicles are becoming a reality. There is still a lot to achieve, but the technology is constantly improving. By building smart digital infrastructures, AI will help reduce the number of accidents and considerably reduce traffic. Also, self-driving delivery trucks and drones will speed up logistic processes, therefore boosting the economy; mostly through one of its bigger engines, the e-commerce industry.

Education

Today, we live in the era of Massive Open Online Courses. Anyone can learn anything online. This is great because the whole world can get access to an education; but it's definitely not enough. A significant improvement would be the personalization of education; everyone learns differently, and at different paces. Some, namely extroverts, will prefer the classroom, while others, introverts, will learn better at home. Some are more visual, while others are more auditory. Taking these and other factors into account, AI is a powerful technology that could deliver personalized training, optimizing everyone's learning curve.

Security

Computer vision has made tremendous technological progress. AI can now detect faces with a high level of accuracy. Not only that, the number of security cameras is increasing significantly. All this could be integrated into a global security platform to reduce crime, increase public safety, and disincentivize people from breaking the law. Besides this, AI and Machine Learning are powerful technologies already used in fraud detection and prevention.

Employment

AI can build powerful recommender systems. We already see platforms of digital recruitment, where AI matches the best candidates to jobs. This not only has a positive impact on the economy, but also on people's happiness, since work makes up more than half of a person's life.

Smart homes and robots

Smart homes, IoT, and connected objects are developing massively. Robots will assist people in their homes, allowing humans to focus on more important activities like their work or spending quality time with their family. They will also help elderly people to live in their home independently, or even allow them to stay active at work, for much longer.

Entertainment and happiness

One downside of technology today is that despite the fact people are so virtually connected, they feel more and more lonely. Loneliness is something we must fight against in this century, as it is very unhealthy for people. AI has a great role to play in this fight, since it is again a powerful recommender system, which can not only recommend relevant movies and songs to users, but also connect people through recommended activities based on their past experiences and common interests.

Through a global smart platform of entertainment, AI technology could help like-minded people to socialize and meet physically instead of virtually.

Another idea to fight loneliness is companion robots, which will be entering homes more and more over the next decade. One branch of AI in the Research and Development phase is emotion creation. This is the branch of AI that will allow robots to show emotions and empathy, and therefore interact more successfully with humans.

Environment

Using computer vision, machines could optimize waste sorting and redistribute the cycles of trash more efficiently. Combining pure AI models with IoT can optimize power and water consumption by individuals. Programs already exist on some platforms that allow people to track their consumption in real time, therefore collecting data. Integrating AI could minimize this consumption, or optimize the distribution cycles for beneficial reuse. Combined with traffic reduction and the development of autonomous vehicles, this will considerably reduce pollution, which will create a healthier environment.

Economy, business, and finance

AI is taking the business world by storm. Earlier, I mentioned the study done by PwC showing how AI could contribute up to $15 trillion to the global economy in 2030 (https://www.pwc.com/gx/en/issues/data-and-analytics/publications/artificial-intelligence-study.html). But how can AI generate so much income? AI can bring significant added value to businesses in three different ways: process automation, profit optimization, and innovation. In my vision of an AI-driven economy, I see the majority of companies adopting at least one AI technology, or having an AI department. In finance, we can already see some jobs being replaced by robots. For example, the number of financial traders was significantly reduced after the development of trading robots that perform well on high-frequency trades.

As you can see, the robot world has a lot of great directions for you to take. AI is already in a dynamic place and it's picking up strong momentum as it moves forward. My professional purpose is to democratize AI and incentivize people to make a positive impact in this world thanks to AI—who knows, perhaps your purpose will be to work with AI for the good of humanity. I'm sure that at least one of these 10 applications resonates in you; if that's the case, work hard to become an AI master and you will have the chance to make a difference.

If you are ready to break into AI, or simply want to increase your knowledge, let's begin!

Summary

In this chapter, you began your AI journey and saw the vast land of opportunities that will open to you. Perhaps you can already think of which industry application might resonate the most in you, so you can become even more passionate about what you do with AI and understand why you're doing it. In the next chapter, you will uncover the AI toolkit you will use in this book.

2
Discover Your AI Toolkit

In the previous chapter, you began your AI journey. Before you continue it, you need your AI toolkit. This book is not just theory; it also contains an easy-to-use toolkit of all the AI models as Python files, ready to run thanks to the amazing Google Colaboratory platform that you will also be introduced to in this chapter.

To fill your AI toolkit, I've prepared a GitHub page containing all the AI implementations for you to download, and Google Colab links of the Python notebooks containing the implementations, all ready to execute via an easy plug and play process.

The GitHub page

You will find all the code for this book ready for you to download from the following GitHub page:

```
https://github.com/PacktPublishing/AI-Crash-Course
```

To download the code, you simply need to click the **Clone or download** button, and then **Download Zip**:

Figure 1: The GitHub repository

Then, once you've downloaded these codes, feel free to open them with your favorite Python **Integrated Development Environment** (**IDE**), whether it's Jupyter Notebook, Spyder, a simple text editor, or even your terminal.

If you've never coded with Python before and have no idea of how to open the files with a Python editor, then no problem; I've prepared the best and simplest solution for you: Colaboratory (or Google Colab).

Colaboratory

Colaboratory is a free and open source environment for Python development that requires no setup and runs entirely on the cloud. It contains all the pre-installed packages required for your AI implementations so that they are ready to run with a simple plug and play process. By plug, I just mean to copy and paste the code inside a new Colab file (I'll explain how to open one next), and by play, I just mean to click on the play button (an example of that follows).

Here is the link to the main page of Colaboratory:

```
https://colab.research.google.com/notebooks/welcome.ipynb
```

You should get a page like this:

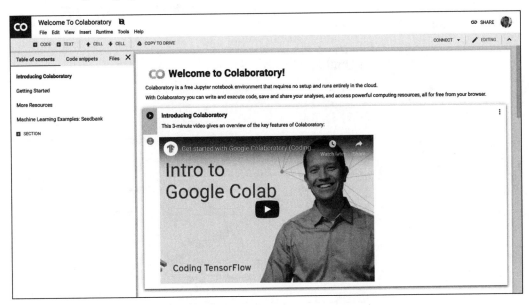

Figure 2: Colaboratory – main page

Click **File** in the upper left, and then click **New Python 3 notebook**:

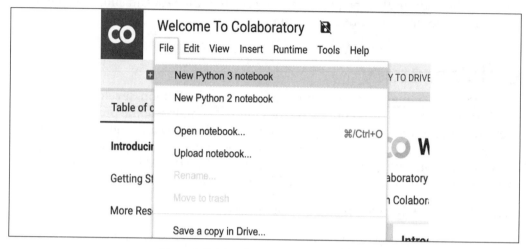

Figure 3: Colaboratory – opening a notebook

Then you will get this view. Paste your Python code inside the cell (red arrow). That's the "plug" part:

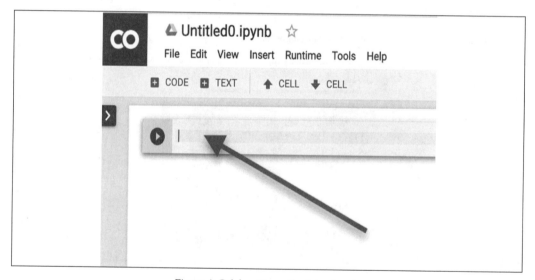

Figure 4: Colaboratory – the "plug" part

I recommend using separate Colaboratory notebooks for each model in this book.

Now let's see the "play" part. Open the Thompson Sampling model in the Chapter 06 folder, implemented inside the thompson_sampling.py file:

Figure 5: GitHub – opening Thompson Sampling

Copy the whole code from inside the Python file; don't worry about understanding the code (or the results) for now. It will all be explained, step by step, in *Chapter 6, AI for Sales and Advertising – Sell like the Wolf of AI Street*:

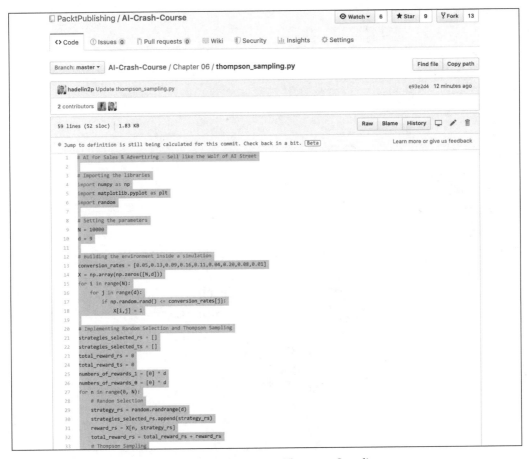

Figure 6: GitHub – copying Thompson Sampling

Next, paste it into Colaboratory (in the cell highlighted by the arrow in *Figure 4*). Then we get this:

```
# AI for Sales & Advertizing - Sell like the Wolf of AI Street

# Importing the libraries
import numpy as np
import matplotlib.pyplot as plt
import random

# Setting the parameters
N = 10000
d = 9

# Building the environment inside a simulation
conversion_rates = [0.05,0.13,0.09,0.16,0.11,0.04,0.20,0.08,0.01]
X = np.array(np.zeros([N,d]))
for i in range(N):
    for j in range(d):
        if np.random.rand() <= conversion_rates[j]:
            X[i,j] = 1

# Implementing Random Selection and Thompson Sampling
strategies_selected_rs = []
strategies_selected_ts = []
total_reward_rs = 0
total_reward_ts = 0
numbers_of_rewards_1 = [0] * d
numbers_of_rewards_0 = [0] * d
for n in range(0, N):
    # Random Selection
    strategy_rs = random.randrange(d)
    strategies_selected_rs.append(strategy_rs)
    reward_rs = X[n, strategy_rs]
    total_reward_rs = total_reward_rs + reward_rs
    # Thompson Sampling
    strategy_ts = 0
    max_random = 0
    for i in range(0, d):
        random_beta = random.betavariate(numbers_of_rewards_1[i] + 1, numbers_of_rewards_0[i] + 1)
        if random_beta > max_random:
            max_random = random_beta
            strategy_ts = i
    reward_ts = X[n, strategy_ts]
    if reward_ts == 1:
        numbers_of_rewards_1[strategy_ts] = numbers_of_rewards_1[strategy_ts] + 1
    else:
        numbers_of_rewards_0[strategy_ts] = numbers_of_rewards_0[strategy_ts] + 1
    strategies_selected_ts.append(strategy_ts)
    total_reward_ts = total_reward_ts + reward_ts

# Computing the Relative Return
relative_return = (total_reward_ts - total_reward_rs) / total_reward_rs * 100
print("Relative Return: {:.0f} %".format(relative_return))

# Plotting the Histogram of Selections
plt.hist(strategies_selected_ts)
plt.title('Histogram of Selections')
plt.xlabel('Strategy')
plt.ylabel('Number of times the strategy was selected')
plt.show()
```

Figure 7: Pasting Thompson Sampling

And now we are ready for the "play" part! Just click the "play" button below:

Figure 8: The "play" part

And the code will execute. Don't pay attention to the result now, as this will all be explained in *Chapter 6, AI for Sales and Advertising – Sell like the Wolf of AI Street*.

You are all set! You now have an AI toolkit that will enable you to follow along with every example in the book.

Before you begin your AI journey in earnest, you must make sure that you have the right basic coding knowledge. This is truly important before becoming a master at AI. If you have little or no experience with Python, make sure that you learn Python in *Chapter 3, Python Fundamentals – Learn How to Code in Python,* as a last preparation phase before you begin exploring the robot world.

Summary

In this chapter, you packed your luggage with our AI toolkit, which included not only the many AI models of this book, but also the very user-friendly Google Colaboratory environment. You saw how easy it was to plug and play our models from GitHub to Colaboratory. Now you just need coding skills to make you ready to begin the real journey. In the next chapter, you will have a chance to learn— or brush up on—your Python fundamentals.

3

Python Fundamentals – Learn How to Code in Python

This chapter is for people who have little or no experience with the Python programming language. If you already know how to use `for`/`while` loops, methods, and classes in Python, you can skip this chapter and you shouldn't have any problems later on.

If, however, you have not used Python before, or have only barely used it, I strongly recommend that you follow this guide. You'll learn how to code the elements of Python I mentioned in the previous paragraph, you'll fully understand the codes included in this book and you'll be able to code in Python on your own. I'll also give you some additional exercises, called "homework" throughout the chapter, which I strongly recommend that you do.

Before you begin, open your Python editor. I recommend using the Google Colab notebook, introduced to you as part of your AI Toolkit in the previous chapter. All the code, along with homework solutions, are provided on the GitHub page of this book in `Chapter 3` in their corresponding section folders. Inside them, you will find two Python files: one (named the same as the section) is the code used in this book, while the `homework.py` file is the solution to the exercise. Instructions for each homework exercise will be provided at the end of each section.

In this chapter, we'll cover the following topics:

- Displaying text
- Variables and operations
- Lists and arrays
- `if` statements and conditions
- `for` and `while` loops

- Functions
- Classes and objects

Especially if you're starting from scratch, cover each section in the order they're presented here, and remember to try your hand at the homework. Let's get started!

Displaying text

We'll begin with the most popular way of introducing any programming language; you'll learn how to display some text in the Python console. The console is a tool that's part of every Python editor, which shows the information we want or displays any errors that occurred (let's hope not to get any!).

The easiest way to show something in our console is to use the print() method, just like this:

```
# Displaying text
print('Hello world!')
```

The text above print, starting with #, is called a comment. Comments are excluded when executing code and are only visible to you.

After running this short code in Google Colab, you'll see this displayed:

```
Hello world!
```

In conclusion, just put what you want to display into the brackets of the print method – text surrounded by quotes, as in this example, or variables.

If you're curious about what variables are, that's great – you'll learn about them after this exercise.

Exercise

Using only one print() method, try to display two or more lines.

Hint: Try using the \n symbol.

The solution is provided in the Chapter 03/Displaying Text/homework.py file on the GitHub page.

Variables and operations

Variables are simply values that are allocated somewhere in the memory of our computer. They are similar to variables in mathematics. They can be anything: text, integers, or floats (a number with precision after the decimal point, such as 2.33).

To create a new variable, you only need to write this:

```
x = 2
```

In this case, we have named a variable x and set its value to 2.

As in mathematics, you can perform some operations on these variables. The most common operations are addition, subtraction, multiplication, and division. The way to write them in Python is like this:

```
x = x + 5     #x += 5

x = x - 3     #x -= 3

x = x * 2.5 #x *= 2.5

x = x / 3     #x /= 3
```

If you look at it for the first time, it doesn't make much sense—how can we write that x = x + 5?

In Python, and in most code, the "=" notation doesn't mean the two terms are equal. It means that we associate the new x value with the value of the old x, plus 5. It is crucial to understand that this is not an equation, but rather the creation of a new variable with the same name as the previous one.

You can also write these operations as shown on the right side, in the comments. You'll usually see them written in this way, since it's more space efficient.

You can also perform these operations on other variables, for example:

```
y = 3
x += y
print(x)
```

Here, we created a new variable y and set it to 3. Then, we added it to our existing x. Also, x will be displayed when you run this code.

So, what does x turn out to be after all these operations? If you run the code, you'll get this:

```
6.333333333333334
```

If you calculate these operations by hand, you will see that x does indeed equal 6.33.

Exercise

Try to find a way to raise one number to the power of another.

Hint: Try using the pow() built-in function for Python.

The solution is provided in the Chapter 03/Variables/homework.py file on the GitHub page.

Lists and arrays

Lists and arrays can be represented with a table. Imagine a **one-dimensional (1D)** vector or a matrix, and you have just imagined a list/array.

Lists and arrays can contain data in them. Data can be anything – variables, other lists or arrays (these are called multi-dimensional lists/arrays), or objects of some classes (we will learn about them later).

For example, this is a 1D list/array containing integers:

	3
	4
	1
	6
	7
	5

And this is an example of a **two-dimensional (2D)** list/array, also containing integers:

	2	9	-5	
	-1	0	4	
	3	1	2	

In order to create a 2D list, you have to create a list of lists. Creating a list is very simple, just like this:

```
L1 = list()
L2 = []

L3 = [3,4,1,6,7,5]
L4 = [[2, 9, -5], [-1, 0, 4], [3, 1, 2]]
```

Here we create four lists: L1, L2, L3 and L4. The first two lists are empty – they have zero elements. The two subsequent lists have some predefined values in them. L3 is a one-dimensional list, same as the one in the first image. L4 is a two-dimensional list, the same as in the second image. As you can see, L4 actually consists of three smaller 1D lists.

Whenever I mention an array, I usually mean a "NumPy" array. NumPy is a Python library (a library is a collection of pre-coded programs that allows you to perform many actions without writing your code from scratch), widely used for list/array operations. You can think of a NumPy array as a special kind of list, with lots of additional functions.

To create a NumPy array, you have to specify a size and use an initialization method. Here's an example:

```
import numpy as np
nparray = np.zeros((5,5))
```

In the first line, we import the NumPy library (as you can see, to import a library, you need to write import) and by using as, we give NumPy the abbreviation np to make it easier to use. Then, we create a new array that we call nparray, which is a 2D array of size 5 x 5, full of zeros. The initialization method is the part after the ".", in this case, we initialize this array as full of zeros, by using the function zeros.

In order to get access to the values in a list or array, you need to give the index of this value. For example, if you wanted to change the first element in the L3 list, you would have to get its index. In Python, indexes start at 0, so you would need to write L3[0]. In fact, you can write print(L3[0]) and execute it, and you will see that, as you might hope, the number 3 will be displayed.

Accessing a single value in a multi-dimensional list/array requires you to input as many indexes as there are dimensions. For example, to get 0 from our L4 list, we would have to write L4[1][1]. L4[1] would return the entire second row, which is a list.

Exercise

Try to find the mean of all the numbers in the L4 list. There are multiple solutions.

Hint: The simplest solution makes use of the NumPy library. Check out some of its functions here: https://docs.scipy.org/doc/numpy/reference/

The solution is provided in the Chapter 03/Lists and Arrays/homework.py file on the GitHub page.

if statements and conditions

Now I would like to introduce you to a very useful tool in programming – if conditions!

They are widely used to check whether a statement is true or not. If the given statement is true, then some instructions for our code are followed.

I'll present this subject to you with some simple code that will tell us whether a number is positive, negative, or equal to 0. The code's very short, so I'll show you all of it at once:

```
a = 5
if a > 0:
    print('a is greater than 0')
elif a == 0:
    print('a is equal to 0')
else:
    print('a is lower than 0')
```

In the first line, we introduce a new variable called a and we give it a value of 5. This is the variable whose value we are going to check.

In the next line we check if this variable is greater than 0. We do this by using an `if` condition. If `a` is greater than 0, then we follow the instructions written in the indented block; in this case, it is only displaying the message `a is greater than 0`.

Then, if the first condition fails, that is, if `a` is lower than or equal to 0, we go to the next condition, which is introduced with `elif` (which is short for `else if`). This statement will check whether `a` is equal to zero or not. If it is, we follow the indented instruction, which will display a message displaying: `a is equal to 0`.

The final condition is introduced via `else`. Instructions included in an `else` condition will always be followed when all other conditions fail. In this case, failing both conditions would mean that a < 0, and therefore we would display `a is lower than 0`.

It's easy to predict what our code will return. It will be the first instruction, `print('a is greater than 0')`. And, in fact, once you run this code, this is what you will get:

```
a is greater than 0
```

In brief, `if` is used to introduce statement checking and the first condition, `elif` is used to check as many further conditions as we want, and `else` is a true statement when all other statements fail.

It's also important to know that once one condition is true, no other conditions are checked. So, in this case, once we enter the first condition and we see that it is true, we no longer check other statements. If you would like to check other conditions, you would need to replace the `elif` and `else` statements with new `if` statements. A new `if` always checks a new condition; therefore, a condition included in an `if` is always checked.

Exercise

Build a condition that will check if a number is divisible by 3 or not.

Hint: You can use a mathematical expression called modulo, which when used, returns the remainder from the division between two numbers. In Python, modulo is represented by `%`. For example:

5 % 3 = 2

71 % 5 = 1

The solution is provided in the `Chapter 03/If Statements/homework.py` file on the GitHub page.

for and while loops

You can think of a loop as continuously repeating the same instructions over and over until some condition is satisfied that breaks this loop. For example, the previous code was not a loop; since it was only executed once, we only checked a once.

There are two types of loops in Python:

- `for` loops
- `while` loops

`for` loops have a specific number of iterations. You can think of an iteration as a single execution of the specific instructions included in the `for` loop. The number of iterations tells the program how many times the instruction inside the loop should be performed.

So, how do you create a for loop? Simply, just like this:

```
for i in range(1, 20):
    print(i)
```

We initialize this loop by writing `for` to specify the type of loop. Then we create a variable `i`, that will be associated with integer values from `range (1,20)`. This means that when we enter this loop for the first time, `i` will be equal to 1, the second time it will be equal to 2, and so on, all the way to 19. Why 19? That's because in Python, upper bounds are excluded, so at the final iteration `i` will be equal to 19. As for our instruction, in this case it's just showing the current `i` in our console by using the `print()` method. It's also important to understand that the main code does not progress until the for loop is finished.

This is what we get once we execute our code:

```
1
2
3
4
5
6
7
8
9
10
11
```

```
12
13
14
15
16
17
18
19
```

You can see that our code displayed every integer higher than 0 and lower than 20.

You can also use a `for` loop to iterate through elements of a list, in the following way:

```
L3 = [3,4,1,6,7,5]
for element in L3:
    print(element)
```

Here we come back to our L3 1D list. This code iterates through every element in the L3 list and displays it. If you run it, you will see all the elements of this array from 3 to 5.

`while` loops, on the other hand, need a condition to stop. They go on as long as the given condition is satisfied. Take this `while` loop, for example:

```
stop = False
i = 0
while stop == False:  # alternatively it can be "while not stop:"
    i += 1
    print(i)
    if i >= 19:
        stop = True
```

Here, we create a new variable called `stop`. This type of variable is called a bool, since it can be assigned only two values – `True` or `False`. Then, we create a variable called `i` that we'll use to count how many times our `while` loop is executed. Next, we create a `while` loop that will go on as long as the variable `stop` is `False`; only once `stop` is changed to `True` will the loop stop.

In the loop, we increase `i` by 1, display it, and check if it is greater or equal to 19. If it is greater or equal to 19, we change `stop` to `True`; and as soon as we change `stop` to `True`, the loop will break!

After executing this code, you will see the exact same output as in the `for` loop example, that is:

```
1
2
3
4
5
6
7
8
9
10
11
12
13
14
15
16
17
18
19
```

It's also very important to know that you can stack `for` and `while` loops inside each other. For example, to display all the elements from the 2D list `L4` we created previously, one after another, you would have to make one `for` loop that iterates through every row, and then another `for` loop (inside the previous one) that iterates through every value in this row. Something like this:

```
L4 = [[2, 9, -5], [-1, 0, 4], [3, 1, 2]]
for row in L4:
    for element in row:
        print(element)
```

And running this yields the following output:

```
2
9
-5
-1
```

0

4

3

1

2

This matches the L4 list.

In conclusion, for and while loops let us perform repetitive tasks with ease. for loops always work on a predefined range; you know exactly when they will stop. while loops work on an undefined range; just by looking at their stop condition, you may not be able to judge how many iterations will happen. while loops work as long as their particular condition is satisfied.

Exercise

Build both for and while loops that can calculate the factorial of a positive integer variable.

Hint: Factorial is a mathematical function that returns the product of all positive integers lower or equal to the argument of this function. This is the equation:

$$f(n) = n * (n - 1) * (n - 2) *...* 1$$

Where:

- *f(n)* – the factorial function
- *n* – the integer in question, the factorial of which we are searching for

This function is represented by ! in mathematics, for example:

5! = 5 * 4 * 3 * 2 * 1 = 120

4! = 4 * 3 * 2 * 1 = 24

The solution is provided in the Chapter 03/For and While Loops/homework.py file on the GitHub page.

Functions

Functions are incredibly useful when you want to increase code readability. You can think of them as blocks of code outside the main flow of code. Functions are executed once they are called in the main code.

You write a function like this:

```
def division(a, b):
    result = a / b
    return result

d = division(3, 5)
print(d)
```

The first three lines are a newly created function called `division`, and the last two lines are part of the main code.

You can create a function by writing `def` and then writing the function's name. After the name, you put brackets and within them write the arguments of the function; these are some variables that you will be able to use inside of your function and are a part of the connection between the main code and the function. In this case, our function takes two arguments: a and b.

Then, once we enter our function, what we do is calculate a divided by b and call this division `result`. Then, in the last line of our function, we say `return` so that when we call this function in code, it will return a value. In this case, the returned value is `result`.

Next, we go back to our main code and call our function. We do that by writing `division` and then in the brackets we input two numbers that we would like to divide. Remember, the `division` function returns a `result` of this division; therefore, we create a variable, d, that will hold this returned value. In the last line, we simply display d to see whether this code really works. If you run it, you'll get the output:

```
0.6
```

As you can confirm by hand, 3 divided by 5 is indeed 0.6; you can test it on other numbers as well.

In real-world code, functions can be much longer, and sometimes even call other functions. You will see them used a lot, even in the other chapters of this book. They also increase code readability, as you will see later; the code I've provided would be impossible to understand without functions.

Exercise

Build a function to calculate the distance between two points on an x,y plane: one with coordinates x1 and y1, and the other with coordinates x2 and y2.

Hint: You can use the following formula:

$$distance = \sqrt{\left(x1-x2\right)^2 + \left(y1-y2\right)^2}$$

The solution is provided in the `Chapter 03/Functions/homework.py` file on the GitHub page.

Classes and objects

Classes, like functions, are another part of code that sits outside of the main code, executed only when called in the main flow of code. Objects are instances of a corresponding class, existing within the main flow of our code. To better understand it, think of a class as a plan of something, for example, a plan of a car. It contains information on how certain components look and work with each other. A class in Python is a general plan of something.

You can think of objects as real-life constructions based on the plan. For example, a real, working, and self-driving car would be an example of an object. You create a plan of a car (which is a class) and then you build a car based on this plan (which is an object). And of course, when you have a plan of something, you can create as many copies as you want; for example, you can run a production line to produce cars.

To give you more insight into classes, we will create a simple bot. We begin with writing a class, like this:

```
class Bot():

    def __init__(self, posx, posy):
        self.posx = posx
        self.posy = posy

    def move(self, speedx, speedy):
        self.posx += speedx
        self.posy += speedy
```

We write `class` to specify that we are creating a new class, which we name `Bot`. Then, a very important step is to write an `__init__()` method, which is a necessity when creating a class. This function is called automatically whenever an object of this class is created in the main flow of the code.

All functions in a class need to take `self` as one argument. So, what is `self`? This parameter specifies that this function and its variables, whose names are preceded by `self`, are a part of this class. We will be able to call the `self` variables once we have an object of this class. Our bot's `__init__()` method also takes two arguments, `posx` and `posy`, which will be the initial position of our bot.

We have also created a method that will move our bot, by increasing or decreasing its `posx` and `posy`. A method is a function tucked inside a class. You can think of it as an instruction on how something has to work when we have a plan. For example, going back to the example of a car, a method could define the way our engine or gearbox works.

Now, you can create an object of this class. Remember, this will be a real-life object, constructed on the basis of a plan (`class`). Before, the class was predefined and didn't work along with your code. After you create an object, the class becomes an integral part of your main code. We can achieve this by doing:

```
bot = Bot(3, 4)
```

This will create a new object of class `Bot`; we called this object `bot`. We also need to specify the two arguments that the `__init__()` method of class `Bot` takes, which are `posx` and `posy`. This isn't optional; when creating an object, you always have to specify all the arguments given in the `__init__()` method.

Now, in the main code, you can move the bot and display its new position, like this:

```
bot.move(2, -1)
print(bot.posx, bot.posy)
```

In the first line, we use the `move` method from our `Bot` class. As you can see in its definition, `move` takes two arguments. These two arguments specify, respectively, by how much we will increase `posx` and `posy`. Then we just display the new `posx` and `posy`. This is where `self` comes into action; if the variables `posx` and `posy` were not preceded by `self` in our `Bot` class, we wouldn't have access to them via the method. Running this code gives us this result:

```
5 3
```

As you can see from the result, our bot moved two units forward on the *x* axis and one unit backward on the *y* axis. Remember, `posx` was set to 3 initially and has now been increased by 2 using the `move` method from the `Bot` class; `posy` was set to 4 initially and has now been decreased by 1, with the use of the same `move` method.

One great advantage of taking the time to code a `Bot` class is that now we are able to create as many bots as we want without making our code any longer. Simply put, objects are copies of a class and we can create as many of them as we want.

In conclusion, you can think of a class as a collection of predefined instructions and closed in methods, and you can think of an object as an instance of this class that is accessible in our code and that runs along with it.

Exercise

Your final challenge will be to build a very simple car class. As arguments, a car object should take the maximum velocity at which the car can move (unit in m/s), as well as the acceleration at which the car is accelerating (unit in m/s²). I also challenge you to build a method that will calculate the time it will take for the car to accelerate from the current speed to the maximum speed, knowing the acceleration (use the current speed as the argument of this method).

Hint: To calculate the time required, you can use the following equation:

$$t = \frac{\left(V_{max} - V_{current} \right)}{a}$$

Where:

- t – time required to achieve the top speed
- V_{max} – maximum speed
- $V_{current}$ – current speed
- a – acceleration

The solution is provided in the `Chapter 03/Classes/homework.py` file on the GitHub page.

Summary

In this chapter, we covered the Python fundamentals that you'll need to keep up with the code presented in this book, from sending a simple text display to the console to writing your very first class in Python. You've now got all the skills you need to continue on your AI journey; in *Chapter 4, AI Foundation Techniques*, we will begin to study the foundational techniques of AI.

4

AI Foundation Techniques

In this chapter, you'll begin your study of AI theory in earnest. You'll start with an introduction to a major branch of AI, called Reinforcement Learning, and the five principles that underpin every Reinforcement Learning model. Those principles will give you the theoretical understanding to make sense of every forthcoming AI model in this book.

What is Reinforcement Learning?

When people refer to AI today, some of them think of Machine Learning, while others think of Reinforcement Learning. I fall into the second category. I always saw Machine Learning as statistical models that have the ability to learn some correlations, from which they make predictions without being explicitly programmed.

While this is, in some way, a form of AI, Machine Learning does not include the process of taking actions and interacting with an environment like we humans do. Indeed, as intelligent human beings, what we constantly keep doing is the following:

1. We observe some input, whether it's what we see with our eyes, what we hear with our ears, or what we remember in our memory.

2. These inputs are then processed in our brain.

3. Eventually, we make decisions and take actions.

This process of interacting with an environment is what we are trying to reproduce in terms of Artificial Intelligence. And to that extent, the branch of AI that works on this is Reinforcement Learning. This is the closest match to the way we think; the most advanced form of Artificial Intelligence, if we see AI as the science that tries to mimic (or surpass) human intelligence.

Reinforcement Learning also has the most impressive results in business applications of AI. For example, Alibaba leveraged Reinforcement Learning to increase its ROI in online advertising by 240% without increasing their advertising budget (see `https://arxiv.org/pdf/1802.09756.pdf`, page 9, Table 1 last row (DCMAB)). We'll tackle the same industry application in this book!

The five principles of Reinforcement Learning

Let's begin building the first pillars of your intuition into how Reinforcement Learning works. These are the fundamental principles of Reinforcement Learning, which will get you started with the right, solid basics in AI.

Here are the five principles:

1. **Principle #1**: The input and output system
2. **Principle #2**: The reward
3. **Principle #3**: The AI environment
4. **Principle #4**: The Markov decision process
5. **Principle #5**: Training and inference

In the following sections, you can read about each one in turn.

Principle #1 – The input and output system

The first step is to understand that today, all AI models are based on the common principle of inputs and outputs. Every single form of Artificial Intelligence, including Machine Learning models, ChatBots, recommender systems, robots, and of course Reinforcement Learning models, will take something as input, and will return another thing as output.

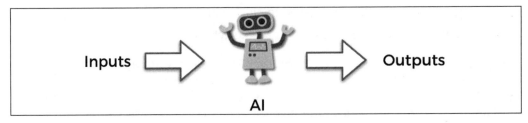

Figure 1: The input and output system

In Reinforcement Learning, these inputs and outputs have a specific name: the input is called the state, or input state. The output is the action performed by the AI. And in the middle, we have nothing other than a function that takes a state as input and returns an action as output. That function is called a policy. Remember the name, "policy," because you will often see it in AI literature.

As an example, consider a self-driving car. Try to imagine what the input and output would be in that case.

The input would be what the embedded computer vision system sees, and the output would be the next move of the car: accelerate, slow down, turn left, turn right, or brake. Note that the output at any time (*t*) could very well be several actions performed at the same time. For instance, the self-driving car can accelerate while at the same time turning left. In the same way, the input at each time (*t*) can be composed of several elements: mainly the image observed by the computer vision system, but also some parameters of the car such as the current speed, the amount of gas remaining in the tank, and so on.

That's the very first important principle in Artificial Intelligence: it is an intelligent system (a policy) that takes some elements as input, does its magic in the middle, and returns some actions to perform as output. Remember that the inputs are also called the **states**.

The next important principle is the reward.

Principle #2 – The reward

Every AI has its performance measured by a reward system. There's nothing confusing about this; the reward is simply a metric that will tell the AI how well it does over time.

The simplest example is a binary reward: 0 or 1. Imagine an AI that has to guess an outcome. If the guess is right, the reward will be 1, and if the guess is wrong, the reward will be 0. This could very well be the reward system defined for an AI; it really can be as simple as that!

A reward doesn't have to be binary, however. It can be continuous. Consider the famous game of *Breakout*:

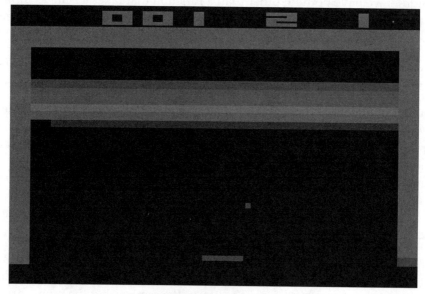

Figure 2: The Breakout game

Imagine an AI playing this game. Try to work out what the reward would be in that case. It could simply be the score; more precisely, the score would be the accumulated reward over time in one game, and the rewards could be defined as the derivative of that score.

This is one of the many ways we could define a reward system for that game. Different AIs will have different reward structures; we will build five rewards systems for five different real-world applications in this book.

With that in mind, remember this as well: the ultimate goal of the AI will always be to maximize the accumulated reward over time.

Those are the first two basic, but fundamental, principles of Artificial Intelligence as it exists today; the input and output system, and the reward. The next thing to consider is the AI environment.

Principle #3 – The AI environment

The third principle is what we call an "AI environment." It is a very simple framework where you define three things at each time (*t*):

- The input (the state)
- The output (the action)
- The reward (the performance metric)

For each and every single AI based on Reinforcement Learning that is built today, we always define an environment composed of the preceding elements. It is, however, important to understand that there are more than these three elements in a given AI environment.

For example, if you are building an AI to beat a car racing game, the environment will also contain the map and the gameplay of that game. Or, in the example of a self-driving car, the environment will also contain all the roads along which the AI is driving and the objects that surround those roads. But what you will always find in common when building any AI, are the three elements of state, action, and reward. The next principle, the Markov decision process, covers how they work in practice.

Principle #4 – The Markov decision process

The Markov decision process, or MDP, is simply a process that models how the AI interacts with the environment over time. The process starts at $t = 0$, and then, at each next iteration, meaning at $t = 1$, $t = 2$, … $t = n$ units of time (where the unit can be anything, for example, 1 second), the AI follows the same format of transition:

1. The AI observes the current state, s_t.
2. The AI performs the action, a_t.
3. The AI receives the reward, $r_t = R(s_t, a_t)$.
4. The AI enters the following state, s_{t+1}.

The goal of the AI is always the same in Reinforcement Learning: it is to maximize the accumulated rewards over time, that is, the sum of all the $r_t = R(s_t, a_t)$ received at each transition.

The following graphic will help you visualize and remember an MDP better, the basis of Reinforcement Learning models:

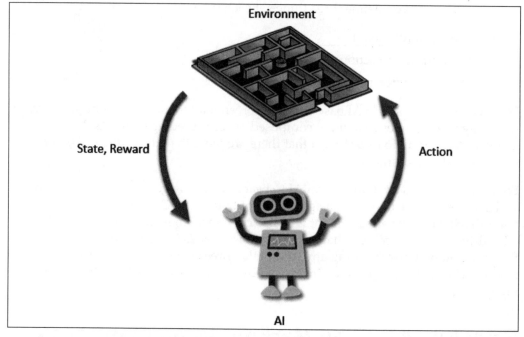

Figure 3: The Markov decision process

Now four essential pillars are already shaping your intuition of AI. Adding a last important one completes the foundation of your understanding of AI. The last principle is training and inference; in training, the AI learns, and in inference, it predicts.

Principle #5 – Training and inference

The final principle you have to understand is the difference between training and inference. When building an AI, there is a time for the training mode, and a separate time for inference mode. I'll explain what that means starting with the training mode.

Training mode

Now you understand, from the three first principles, that the very first step of building an AI is to build an environment in which the input states, the output actions, and a system of rewards are clearly defined. From the fourth principle, you also understand that inside this environment we will build an AI to interact with it, trying to maximize the total reward accumulated over time.

To put it simply, there will be a preliminary (and long) period of time during which the AI will be trained to do that. That period of time is called the training; we can also say that the AI is in training mode. During that time, the AI tries to accomplish a certain goal over and over again until it succeeds. After each attempt, the parameters of the AI model are modified in order to do better at the next attempt.

For example, let's say you're building a self-driving car and you want it to go from point *A* to point *B*. Let's also imagine that there are some obstacles that you want your self-driving car to avoid. Here is how the training process happens:

1. You choose an AI model, which can be Thompson Sampling (*Chapters 5 and 6*), Q-learning (*Chapters 7 and 8*), deep Q-learning (*Chapters 9, 10, and 11*) or even deep convolutional Q-learning (*Chapters 12 and 13*).

2. You initialize the parameters of the model.

3. Your AI tries to go from *A* to *B* (by observing the states and performing its actions). During this first attempt, the closer it gets to *B*, the higher reward you give to the AI. If it fails reaching *B* or hits an obstacle, you give the AI a very bad reward. If it manages to reach *B* without hitting any obstacle, you give the AI an extremely good reward. It's just like you would train a dog to sit: you give the dog a treat or say "good boy" (positive reward) if the dog sits. And you give the dog whatever small punishment you need to if the dog disobeys (negative reward). That process is training, and it works the same way in Reinforcement Learning.

4. At the end of the attempt (also called an episode), you modify the parameters of the model in order to do better next time. The parameters are modified intelligently, either iteratively through equations (Q-Learning), or by using Machine Learning and Deep Learning techniques such as stochastic gradient descent or backpropagation. All these techniques will be covered in this book.

5. You repeat steps 3 and 4 again, and again, until you reach the desired performance; that is, until you have your fully non-dangerous autonomous car!

So, that's training. Now, how about inference?

Inference mode

Inference mode simply comes after your AI is fully trained and ready to perform well. It will simply consist of interacting with the environment by performing the actions to accomplish the goal the AI was trained to achieve before in training mode. In inference mode, no parameters are modified at the end of each episode.

For example, imagine you have an AI company that builds customized AI solutions for businesses, and one of your clients asked you to build an AI to optimize the flows in a smart grid. First, you'd enter an R&D phase during which you would train your AI to optimize these flows (training mode), and as soon as you reached a good level of performance, you'd deliver your AI to your client and go into production. Your AI would regulate the flows in the smart grid only by observing the current states of the grid and performing the actions it has been trained to do. That's inference mode.

Sometimes, the environment is subject to change, in which case you have to alternate fast between training and inference modes so that your AI can adapt to the new changes in the environment. An even better solution is to train your AI model every day, and go into inference mode with the most recently trained model.

That was the last fundamental principle common to every AI. Congratulations – now you already have a solid basic understanding of Artificial Intelligence! Since you have that, you are ready to tackle your very first AI model in the next chapter: a simple yet very powerful one, still widely used today in business and marketing, to solve a problem that has the delightful name of the multi-armed bandit problem.

Summary

In this chapter, you learned the five fundamental principles of Artificial Intelligence from a Reinforcement Learning perspective. Firstly, an AI is a system that takes an observation (values, images, or any data) as input, and returns an action to perform as output (principle #1). Then, there is a reward system that helps it measure its performance. The AI will learn through trial and error based on the reward it gets over time (principle #2). The input (state), the output (action), and the reward system define the AI environment (principle #3). The AI interacts with this environment through the Markov decision process (principle #4). Finally, in training mode, the AI learns how to maximize its total reward by updating its parameters through the iterations, and in inference mode, the AI simply performs its actions over full episodes without updating any of its parameters – that is to say, without learning (principle #5).

In the next chapter, you will learn about Thompson Sampling, a simple Reinforcement Learning model, and use it to solve the multi-armed bandit problem.

5

Your First AI Model –
Beware the Bandits!

In this chapter, you'll get to grips with your very first AI model! You're going to make a model that will solve the very well-known multi-armed bandit problem. This is a classic problem in AI, and it's also widely encountered in many real-world business problems.

The multi-armed bandit problem

Imagine you are in Las Vegas, in your favorite casino. You are in a room containing five slot machines. For each of them the game is the same: you bet a certain amount of money, say 1 dollar, you pull the arm, and then the machine will either take your money, or give you twice your money back. Remember the rewards we talked about in the previous chapter? Let's say that if the machine takes your money, your reward is -1, and if the machine returns you twice your money, your reward is +1.

As you can see, you're already starting to define an AI environment, which I'll remind you is absolutely fundamental when solving a problem with AI. So far, the AI isn't there, but it will come soon. You always start by defining the environment.

You've defined the rewards; you'll define the states (inputs) and actions (outputs) later. Now, still in the process of defining the environment, let's say that you know, somehow, that one of these machines has a higher probability of giving you a +1 reward than the others when you pull its arm. It doesn't matter how you know this info, but it must be part of the problem assumptions. Rest assured, this assumption is always naturally verified in the real-world business problems mentioned above where the multi-armed bandit problem can be applied.

Your goal, as in any AI environment, is to obtain the highest accumulated reward during your time of play. Let's say you are going to bet 1,000 dollars in total, meaning that you are going to bet 1 dollar, 1,000 times, each time by pulling the arm of any of these five slot machines. The question is:

What should be your strategy, so that after having played 1,000 times, you get the maximum amount of money to take home with you?

The first step of your strategy must be to figure out, in the minimum number of plays, which of these five slot machines has the highest chance of giving you a 1 reward. In other words, you have to quickly figure out the slot machine with the highest success rate. Then, as soon as you figure it out, you simply need to keep playing on that most successful slot machine.

Finding the most successful slot machine is not hard; one simple strategy could be to play 100 times on each of these five slot machines and then, at the end, look at which of them gave you more money. Statistically, this gives you a good chance of finding that most generous slot machine.

All the challenge is in "quickly". The hardest part is to find the best slot machine **in a minimum number of trials**. This is where your first AI model comes into play.

The Thompson Sampling model

You're going to build this model straight away. Right now, you'll build a simple implementation of this method, and later you will be shown the theory behind it. Let's get right into it!

As we defined previously, our problem is trying to find the best slot machine with the highest winning chance out of many. A not-so-optimal solution would be to play 100 rounds on each of our slot machines and see which one has the highest winning rate. A better solution is a method called Thompson Sampling.

I won't go too deeply into the theory behind it; we'll cover that later. For now, it is enough to say that Thompson Sampling uses a distribution function (distributions will be explained further in this chapter), called Beta, that takes two arguments. For simplicity's sake, let's say that the higher the first argument is, the better our slot machine is, and the higher the second argument is, the worse our slot machine is.

Therefore, we can define this function as:

$$x = \beta(a, b)$$

where:

- x – a random choice from our Beta distribution
- β – our Beta function
- a – the first argument
- b – the second argument

Don't worry if you don't understand this entirely quite yet; you'll read all about it later.

Coding the model

Let's start coding our solution. All this code is also available on the GitHub page of this book in the Chapter 05 folder. Here we go with the first code section:

```
# Importing the libraries
import numpy as np
```

You'll only need one library, called NumPy. This is a very useful library, helping when we are dealing with multi-dimensional arrays and lists in general. Give it the abbreviation np, which is the industry standard, so that it will be easier to use.

Now we have to understand something very important. You are creating a simulation whose aim is to simulate real-life situations. In reality, every slot machine gives us some chance of winning, and some machines have it higher than others. Therefore, when simulating this environment, you have to do the same thing. It is important to remember, however, that our AI will not know these predefined winning rates. It cannot just read them and judge, based on these rates, which machine is the best.

For this example, let's call this list of winning chances conversionRates.

```
# Setting conversion rates and the number of samples
conversionRates = [0.15, 0.04, 0.13, 0.11, 0.05]
N = 10000
d = len(conversionRates)
```

Here, you have five slot machines. They have some win chance; for example, slot machine no. 1 offers a 15% chance of a win. Then you create a number of samples, N. Remember, you are performing a simulation, so you need to have a predefined dataset that will tell you whether you won or not when you're playing. You also introduce a variable, d, which is the length of your conversion rates list; that is, the number of slot machines. It's useful to use short variable names like that, because the code would be longer and less readable otherwise.

Do you have an idea of what you should do next? You are running a simulation, so you need to have a predefined set of wins and losses for every slot machine for every sample. I highly recommend that you try to do this on your own. You need to have a set that will tell you if at some timestep i you have won or not by playing a certain slot machine. The answer is in the next snippet of code.

```
# Creating the dataset
X = np.zeros((N, d))
for i in range(N):
    for j in range(d):
        if np.random.rand() < conversionRates[j]:
            X[i][j] = 1
```

In the first line, you create a 2d-array full of zeros, of size N * d. This means that you've created an array with N (in this case 10000) rows and d (in this case 5) columns. Then, in a for loop, you iterate through every row in that 2d-array X. In a nested for loop, you iterate through each column in that row. In line 5 of the preceding code snippet, for each slot machine (each column), we check if a random float number from range (0,1) is smaller than the conversion rate for the corresponding slot machine.

That's just like playing the slot machine; since there is an equal chance of getting any float number from this range, the chances of getting a number smaller than x (where x is also in range (0,1)) is equal to x. For example, for d = 0.15, there are 15 instances out of 100 of getting a smaller float number than 0.15, and thus a 15% chance of returning a high reward for slot machine 1. In other words, if the random float is smaller, then that means you will win if you play this certain machine at this certain timestep.

To make sure you understand, if one of the N samples from your dataset X looks like this: [0, 1, 0, 0, 1], you would win at that point in time by playing slot machine no. 2 or no. 5.

Next, you need to create two arrays that will count how many times you have lost and won by playing each slot machine, like this:

```
# Making arrays to count our losses and wins
nPosReward = np.zeros(d)
nNegReward = np.zeros(d)
```

Name them `nPosReward` (number of wins) and `nNegReward` (number of losses).

Now that you have made a simulation set and these two counters, you can start coding some Thompson Sampling. Keep in mind that the theory, as well as another example, will be covered later.

Next, initialize a `for` loop that will iterate through every sample in our dataset and choose the best slot machine. Initially, only create two variables, one called `selected`, which will tell you which slot machine was chosen, and `maxRandom`, which you will use to get the highest Beta distribution guess across all slot machines:

```
# Taking our best slot machine through beta distribution and updating
its losses and wins
for i in range(N):
    selected = 0
    maxRandom = 0
```

So now you can get to the core of Thompson Sampling. You'll take random guesses from our Beta distribution and find the highest value across all your slot machines.

You can use a method taken from NumPy, called `np.random.beta(a,b)`, that returns this random guess. Knowing that, try to find the highest guess and the best machine on your own! It is totally fine if you fail—we haven't covered the theory yet—and I will provide you with an answer. Good luck!

I hope you've given it a try. Whether it's worked out for you or not, here's my answer:

```
    for j in range(d):
        randomBeta = np.random.beta(nPosReward[j] + 1, nNegReward[j] +
1)
        if randomBeta > maxRandom:
            maxRandom = randomBeta
            selected = j
```

You haven't missed anything—this is all the code needed for this task. You create a `for` loop to iterate through every slot machine and find the best one. For each slot machine of index `j` (remember that you are still in the bigger `for` loop with index `i`), you take a random draw, called `randomBeta`, from our Beta distribution, and check if it is greater than `maxRandom`.

If it is, then you reassign `maxRandom` to be equal to `randomBeta`, and set `selected` to be equal to the index of this new highest-guess slot machine `j`. It is also worth mentioning what the `a` and `b` arguments of the Beta function are in this case; they're the number of wins and losses we've had on the specific slot machine. Remember, the bigger the first argument, the better, and the higher our random guess will be; the bigger the second argument, the worse, and the lower our random guess will be.

Now that you have selected the best slot machine, what do you think you should do next?

You have to update your `nPosReward` or `nNegReward` depending on whether you have won or not. We can do that with this code:

```
if X[i][selected] == 1:
    nPosReward[selected] += 1
else:
    nNegReward[selected] += 1
```

Here, you can see the use of the `X` array you created earlier. You check if you have won this round by checking if there's a `1` in the appropriate place in your `X` array. If you win, you update the index corresponding to the selected machine in `nPosReward` by adding 1. If you lose, however, you update `nNegReward` by adding 1 in the same index there. You can clearly see that if you win, next time, your random guess from the Beta distribution for that machine will be higher; and if you lose, it will be lower.

This code works already, although it is worth adding a few lines of code to display which slot machine your code considers the best:

```
# Showing which slot machine is considered the best
nSelected = nPosReward + nNegReward
for i in range(d):
    print('Machine number ' + str(i + 1) + ' was selected ' +
str(nSelected[i]) + ' times')
print('Conclusion: Best machine is machine number ' + str(np.
argmax(nSelected) + 1))
```

Here, you simply display how many times each slot machine was chosen by your algorithm. To get these numbers you can add together the lists `nPosReward` and `nNegReward`. In the final line, you show which machine was chosen the highest number of times, making it the slot machine that is considered the best.

Now, you can just run your code and see the results:

```
Machine number 1 was selected 7927.0 times
Machine number 2 was selected 82.0 times
Machine number 3 was selected 1622.0 times
Machine number 4 was selected 306.0 times
Machine number 5 was selected 63.0 times
Conclusion: Best machine is machine number 1
```

As we can see, your algorithm **quickly** found out that machine no. 1 is the best. It did it in around 2,000 rounds (2,000 samples in your x set).

Understanding the model

Thompson Sampling is, by far, the best model for this kind of problem; at the end of this chapter, you will see a comparison with another method. Here's how it works its magic. The first thing we do, when finding the best slot machine, is obviously to play the arm of each of the five slot machines one by one. So here we go:

Round 1: We play the arm of slot machine number 1. Let's say we get reward 0.

Round 2: We play the arm of slot machine number 2. Let's say we get reward 1.

Round 3: We play the arm of slot machine number 3. Let's say we get reward 0.

Round 4: We play the arm of slot machine number 4. Let's say we get reward 0.

Round 5: We play the arm of slot machine number 5. Let's say we get reward 1.

Now, why do you think we had to do this? We only did that to collect some starting information from each of the slot machines. This information will be needed in future rounds.

Now, things start to get interesting. What are we going to do at round 6? Which arm are we going to play?

Well, we need to look back at what happened during the first five rounds. For each slot machine, we introduce two new variables, one that counts the number of times the slot machine returned a 0 reward, and another one that counts the number of times the slot machine returned a 1 reward.

Let's denote these variables as $N_i^0(n)$ and $N_i^1(n)$, where $N_i^0(n)$ is the number of times slot machine number i returned reward 0 up to round n, and $N_i^1(n)$ is the number of times slot machine number i returned reward 1 up to round n. These two variables are denoted by nNegReward and nPosReward in our code. So, based on what we've obtained so far at round 5, let's give some values examples of these variables:

$N_1^0(1)=1$ means that slot machine 1 has returned 1 loss over 1 round.

$N_1^1(1)=0$ means that slot machine 1 has returned 0 wins over 1 round.

$N_2^0(1)=0$ means that slot machine 2 has returned 0 losses over 1 round.

$N_2^1(1)=1$ means that slot machine 2 has returned 1 win over 1 round.

$N_5^0(4)=0$ means that slot machine 5 has returned 0 losses over 4 rounds.

$N_5^1(4)=0$ means that slot machine 5 has returned 0 wins over 4 rounds.

$N_5^0(5)=0$ means that slot machine 5 has returned 0 losses over 5 rounds.

$N_5^1(5)=1$ means that slot machine 5 has returned 1 win over 5 rounds.

Alright, that was the easy part. The good news is that we've created all the variables we needed for our AI. The bad news is that now comes the hard part, the math. If you think math is good news, I like your spirit; but don't worry if you don't like math, I won't let you down.

What is a distribution?

The next step of our AI journey is to introduce distributions in mathematics. For this, I'll give you a simple definition with my own words, not the very formal ones you find in math books. I want to make sure everybody understands. Here it is: the distribution of a variable is a function that will give, for each value in the possible range of values the variable could take, the probability that this variable is equal to that value.

Let's really understand what it is through an example:

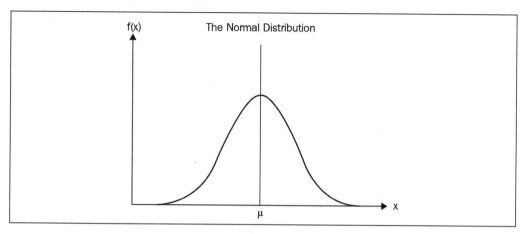

Figure 1: The normal distribution

In the preceding graph, you can see an example of a distribution. Now, remember in the definition I gave you, I mentioned two measures: "range of values the variable could take", and "probability that this variable is equal to that value". In any distribution, on the x-axis you have the range of values the variable could take, and on the y-axis you have the probability that the variable is equal to each value.

Don't worry if this isn't clear yet. To extend our example, let's say that on the preceding graph, this variable is the annual salary people have in a specific country.

On the x-axis, we would have the range of annual salaries from the minimum wage to the maximum wage, let's say from 15,000 dollars to 150,000 dollars. And on the y-axis, we would have the probabilities that a person would have that salary.

Now it should make more sense. For the low salaries, the curve is low, meaning that the probability that an individual earns a salary of around 15,000 dollars is low.

Then, up to the center of the x-axis, marked as μ, which is the average of the salaries, the probabilities of people's salaries increase. Let's say that μ is equal to 45,000 dollars. We intuitively understand that the probability that an individual in a specific country earns 45,000 dollars per year is the highest, simply because the majority of people earn something in the region of 45,000 dollars per year. And that's exactly why the distribution in the graph is the highest at this salary.

The higher we go above an annual salary of 45,000 dollars, the fewer people we'll find earning such salaries, and therefore the probability of people earning such salaries will decrease, until we go beyond an annual salary of 150,000 dollars, where very few people earn that much, therefore leading to a close-to-zero probability.

Alright, that was the distribution explained intuitively. Now, you have to know that there are many types of distributions: Gaussian distributions (that look like the preceding graph), normal distributions (Gaussian distribution of mean 0 and variance 1), Beta distributions, and many more.

That's the next step: **Beta distributions**. The Beta distribution is at the heart of the AI we built to solve our bandit problem. Here are what Beta distributions look like:

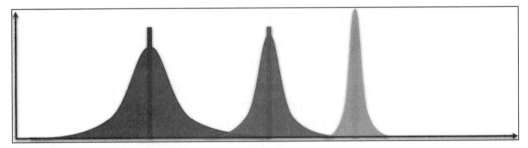

Figure 2: Three Beta distributions

Let's do some practice to make sure you understand how distributions work. Imagine these three distributions correspond to three different countries, and again let's say that they are the distributions of salaries in these countries. Which country has the highest salaries? Is it the purple one, the green one, or the yellow one? The answer is the yellow one, of course! It is in this country that we have positive probabilities for the highest salaries (remember, the salaries are on the x-axis, and the probabilities are on the y-axis).

That was just a quick test to make sure you were with me. Now, you don't have to remember the exact formula of a Beta distribution, but you do have to know that it has two parameters and how they impact the distribution. Don't forget that this was already mentioned when we solved the problem in practice, now it is explained in much more detail.

If we denote these two parameters as a and b again, we can denote the Beta distribution with the following:

$$y = \beta(x, a, b)$$

You might be asking what just happened—Why did x appear? Don't worry, we will demystify all this. In the formula above, y is the probability, β is a function of x only, x is the salary, and a, b are the two parameters present in any Beta distribution. Again, you don't have to know the exact definition of the function β, but just keep in mind the shape of its curve as given in the preceding graph.

However, what is really important for you to understand now is the role of the two parameters a and b. Following are the two points that you must know and visualize in your head:

1. Given two Beta distributions with the same parameter b, the one having a larger parameter a will be shifted more to the right.

2. Given two Beta distributions with the same parameter a, the one having a larger parameter b will be shifted more to the left.

That's it! That's enough to have an intuitive understanding of how our AI will solve the Bandit problem. In other words, the larger the parameter a, the more it will shift the Beta distribution to the right, and the larger the parameter b, the more it will shift the Beta distribution to the left.

Let's practice this! If I give you the following three Beta distributions:

1. $\beta(1,5)$
2. $\beta(5,1)$
3. $\beta(3,3)$

Could you tell me which of the three Beta distributions in the following graph they would approximately look like?

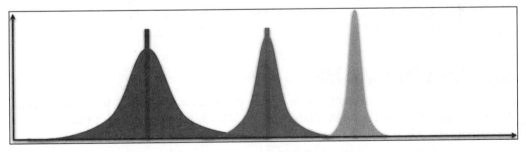

Figure 3: Three Beta distributions

Based on the two statements above, $\beta(1,5)$ is the purple one, $\beta(5,1)$ is the yellow one, and $\beta(3,3)$ is the green one. Congratulations to you if you guessed that right!

Now you are ready to solve our bandit problem. But let me ask you a question first, which might lead you to understand the magic faster than this book:

If, instead of the salaries in a country, the x-axis contained the success rates of the machines in the casino, and if each of the three Beta distributions represented one particular slot machine, which one would you choose to bet your 1,000 dollars?

You would choose the yellow one!

Of course! This distribution has positive probabilities for the highest conversion rates, since it is the one most shifted to the right.

This was already discussed in the previous code section of this chapter; I told you there that the higher the first parameter, the better the slot machine. Indeed, the Beta distribution will be shifted more to the right, meaning that this slot machine has a higher chance of giving us a win. Additionally, the higher the second parameter, the worse the slot machine is and now, the Beta distribution will be shifted to the left, meaning that this machine has a lower chance of us winning.

And now another question, before we solve our bandit problem. Remembering that you have five slot machines to play with, try to answer this question: if the five slot machines are associated with the following five Beta distributions of success rates:

$\beta(1,3)$, $\beta(1,5)$, $\beta(3,3)$, $\beta(5,3)$, and $\beta(5,1)$,

Which one would you pick to bet your 1,000 dollars?

The answer is $\beta(5,1)$!

Of course, again! Because it is the one with the largest parameter *a* and the lowest parameter *b*, therefore the most shifted to the right, and hence the one having the positive probabilities for the highest conversion rates.

If you are still with me, you are definitely ready to understand the AI magic. If not, please read through this section again. In the next section, I will finally reveal what happens next after Round 5.

Tackling the MABP

What we are going to do from now on before playing each round is to associate each slot machine with a specific Beta distribution. At each round *n*, the slot machine number *i* (*i*=1,2,3,4,5) will be associated with the following Beta distribution:

$$\beta\left(N_i^1(n)+1, N_i^0(n)+1\right)$$

Here, you should recall the following:

- $N_i^1(n)$ is the number of times the slot machine number *i* returned a 1 reward up to round *n*.

- $N_i^0(n)$ is the number of times the slot machine number *i* returned a 0 reward up to round *n*.

Remember, in the Beta distribution $\beta(a,b)$, the higher the parameter *a*, the more that shifts the distribution to the right. The higher the parameter *b*, the more that shifts the distribution to the left. Therefore, since at each round *n* and for each slot machine, the parameter *a* is the number of times (plus 1) it returned 1 up to round *n*, and the parameter *b* is the number of times (plus 1) it returned 0 up to round *n*, then that means the following: the more the slot machine returns 1 (success), the more its distribution will be shifted to the right; and the more the slot machine returns 0 (failure), the more its distribution will be shifted to the left.

Congratulations if you figured out what *a* and *b* should be on your own. We already used them in the practical tutorial above; we had two arrays, nPosReward and nNegReward, that correspond to $N_i^1(n)$ and $N_i^0(n)$ respectively.

Once you understand this, try to figure out the strategy before I give you the solution.

Alright, you are about to see the magic. What we are going to do, before playing the arm at each round, is take a random draw from each of the five distributions corresponding to the five slot machines. In case you're not clear what that means, I'll explain. Let me show you again the graph of the three Beta distributions:

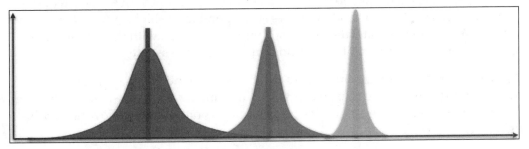

Figure 4: Three Beta distributions

What did I mean by taking a random draw? First, remember that for our bandit problem, on the *x*-axis, we have the success rates from 0 to 1. For example, $x = 0.25$ means that the machine returns a 1 reward (success) 25% of the time. Then, on the *y*-axis, we still have the probabilities to have these success rates.

Let's focus on one distribution, for example, the purple one. What would it mean to take a random draw from that distribution? That would mean very simply that we randomly pick a value on the x-axis where the distribution is positive, such that the x values where the probability is the highest will get the highest chance to be picked. For example, let's say the top of the purple curve corresponds to $x = 0.2$ and $y = 0.35$.

Then, taking a random draw from that purple distribution means that we will have a 35% chance to pick a success rate of 20%. To generalize this, let's say that $y = \beta_{purple}(x)$ is the function associated with the purple distribution, so taking a random draw from that purple distribution means that for each success rate x on the x-axis, we will have $\beta_{purple}(x)$ chance of picking x. That is what "to take a random draw from a distribution" means, and this is also called "to sample a distribution".

Now that you understand this, let's see where we left off. We said that before playing the arm at each round, we were going to take a random draw from each of the five distributions corresponding to the five slot machines. We thus obtain five values on the x-axis, each one corresponding to each of the five slot machines. Then, here comes the crucial question, the one that will tell whether you have the right intuition about the strategy.

According to you, which slot machine are you going to play, based on the observation of these five values? I really want you to take some time to answer this question, because right now, we are at the heart of the strategy (you can also have a look at our previously written code). The answer can be found in the next paragraph.

I really hope you tried figuring this out by yourself: the slot machine that you are going to play next is the one for which we got the highest of the five random draws. Why? Because the highest random draws correspond to the highest success rate, and for this highest success rate, the Beta distribution associated with the slot machine picked has positive probabilities around that highest success rate.

Since we want to maximize the success rate of the machines we play (because we want to make money), we must pick the slot machine for which the Beta distribution has positive probabilities around the highest success rates. In the following graph, that's the yellow distribution.

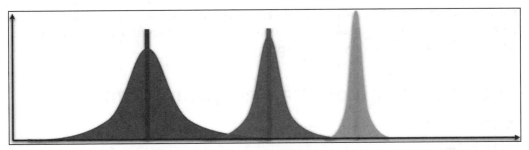

Figure 5: Three Beta distributions

Now, we must take a step back. I've been in your situation many times when I am learning something new and technical, which sometimes felt overwhelming. In that case, the best move is to take a step back, which is exactly what we are going to do now by giving a recap of the strategy and its intuition.

The Thompson Sampling strategy in three steps

After we play each of the five slot machines over the first five rounds, here's what the AI will do at each round n:

1. For each slot machine i ($i=1,2,3,4,5$), we take a random draw $\theta_i(n)$ from its Beta distribution:

   $$\theta_i(n) \sim \beta\left(N_i^1(n)+1, N_i^0(n)+1\right)$$

   where:

   $N_i^1(n)$ is the number of times the slot machine number i returned a 1 reward up to round n.

   $N_i^0(n)$ is the number of times the slot machine number i returned a 0 reward up to round n.

2. We pull the arm of the slot machine $s(n)$ that has the highest sampled $\theta_i(n)$:

   $$s(n) = \underset{i=1,2,3,4,5}{\operatorname{argmax}}\left(\theta_i(n)\right)$$

3. We don't forget to update $N_{s(n)}^1(n)$ or $N_{s(n)}^0(n)$:

   If the played slot machine $s(n)$ returned a 1 reward:

   $$N_{s(n)}^1(n) := N_{s(n)}^1(n)+1$$

   If the played slot machine $s(n)$ returned a 0 reward:

   $$N_{s(n)}^0(n) := N_{s(n)}^0(n)+1$$

Then, we repeat these three steps at each round until we spend our 1,000 dollars. This strategy, called Thompson Sampling, is a basic but powerful model of a specific branch of AI, called Reinforcement Learning.

The final touch of shaping your Thompson Sampling
intuition

Your intuition about why and how this works should be as follows (try to keep it in mind or visualize it on the graphic):

Each slot machine has its own Beta distribution. Over the rounds, the Beta distribution of the slot machine with the highest conversion rate will be progressively shifted to the right, and the Beta distributions of the strategies with lower conversion rates will be progressively shifted to the left (Steps 1 and 3). Therefore, because of Step 2, the slot machine with the highest conversion rate will be selected more and more.

And voilà! Congratulations—you just learned about a powerful AI model, a massive step in your journey. To see Thompson Sampling in action and check that it indeed works, I won't force you to go to a casino and try it out; We'll apply it to another real-life model in *Chapter 6, AI for Sales and Advertising – Sell like the Wolf of AI Street*.

Finally, let me finish this theory tutorial with a question for you. Remember earlier in the book I told you that any AI we build today takes as input a state, returns as output an action to play, and, after playing the action, gets a reward (positive or negative). **For this particular bandit problem, what are the input states, the actions played, and the rewards received?** Think about this before reading the next paragraph.

Here we go with the answer:

- The input state is the exact round we've reached, including the information of the two parameters $N_{s(n)}^1(n)$ and $N_{s(n)}^0(n)$.
- The output action is the arm we pull from the selected slot machine.
- The reward is 1 or 0, 1 if the slot machine returns twice our dollar invested, and 0 if we lose our dollar.

Congratulations to you if you answered that one correctly, and for tackling this first AI model, Thompson Sampling. And don't forget, in *Chapter 6, AI for Sales and Advertising – Sell like the Wolf of AI Street*, we put this into practice to solve a real-world business problem.

Thompson Sampling against the standard model

When I learned Thompson Sampling for the first time, I had one main question in my mind: is it really that good? In fact, if you were to run the standard model (by "standard model" I mean playing every slot machine a certain number of times) and Thompson Sampling separately you might not see much difference; you would likely come to the conclusion that they work pretty much as well as each other.

To check whether it is true that Thompson Sampling isn't any better, I implemented a code to test both solutions on many different scenarios. The changes included: number of samples (200 or 1,000 or 5,000), number of slot machines (from 3 to 20), and conversion rate ranges (ranges in which conversion rates could be set: 0-0.1; 0-0.3; 0-0.5).

Every scenario was tested 100 times to compute the accuracy of each model.

The results and the code used are provided in the resultsModified.xlsx and comparison.py files, respectively, in Chapter 05 of this book's GitHub page. Here, you can see some graphs taken from this Excel file that show the performance of both models:

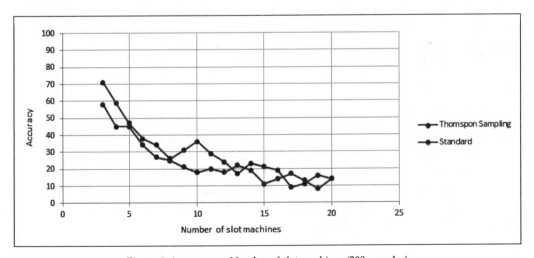

Figure 6: Accuracy vs. Number of slot machines (200 samples)

This first graph in *Figure 6* illustrates the accuracy of both models depending on the number of slot machines. The number of samples was set to 200 and the conversion rate ranges were set to 0-0.1, meaning that the differences between these rates were minor. This is the toughest setting for this comparison. Overall, Thompson Sampling performed better than the standard model (22% better).

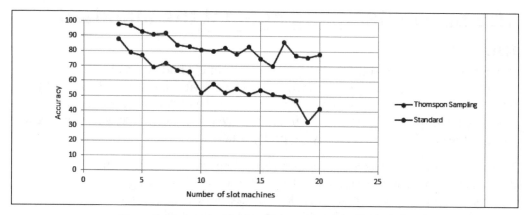

Figure 7: Accuracy vs. Number of slot machines (5,000 samples)

This second graph in *Figure 7* shows the performance under the easiest conditions. The number of samples was set to 5,000 and the conversion rate ranges were set to 0-0.5, meaning that the differences were clearly visible. The overall drop of accuracy for Thompson Sampling is smaller than the drop in accuracy for the standard solution. Thompson Sampling performed significantly better this time (41% better).

Taking all scenarios into consideration, Thompson Sampling achieved a mean accuracy of 57% and the standard model achieved 43% accuracy. This is a significant difference taking into account the fact that very tough scenarios were tested (for example, only 200 samples, a range of 0-0.1, and 20 slot machines).

Summary

Thompson Sampling is a powerful sampling technique that enables you to quickly figure out the highest of a number of unknown conversion rates. It is always applied in the same frame, called the multi-armed bandit problem, which in the classic sense is composed of several slot machines, each one having a different conversion rate of positive outcomes. We had a first glance at how this AI solves this problem better and faster than standard methods.

In the next chapter, we will perform a full practical activity where we will see how the multi-armed bandit frame can easily model a business problem — online advertising — and how Thompson Sampling can bring significant added value.

6

AI for Sales and Advertising – Sell like the Wolf of AI Street

Now it's time to put your new skills into practice, start coding, and shape up your AI skills! You've learned all about Thompson Sampling, and now it's time to implement this AI model to solve a real-world problem, maximizing the sales of an e-commerce business.

In this practical exercise, you'll really take action and build the AI yourself to solve the problem. It's really important that you stay active in this chapter, because this is where you will have the chance to learn by doing, which is the most effective way to learn something; practice truly makes perfect. In other words, I want you to be the hero of this AI adventure. You, and not me. Ready?

Problem to solve

Imagine an e-commerce business that has millions of customers. These customers are people buying products on the website from time to time, getting those products delivered to their homes. The business is doing well, but the board of executives has decided to follow an action plan to maximize revenue.

This plan consists of offering the customers the option to subscribe to a premium plan, which will give them benefits like reduced prices, special deals, and so on. This premium plan is offered at a yearly price of $200, and the goal of this e-commerce business is, of course, to get the maximum number of customers to subscribe to this premium plan. Let's do some quick math to give us some motivation for building an AI to maximize the revenue of this business.

Let's say that this e-commerce business has 100 million customers. Now consider two strategies to convert the customers to the premium plan: a bad one, with a conversion rate of 1%, and a good one, with a conversion rate of 11%. If the business deploys the bad strategy, in one year it will make a total of: 100,000,000 × 0.01 × 200 = $200,000,000 in extra revenue from the premium plan subscriptions.

On the other hand, if the business deploys the good strategy, in one year it will make a total of: 100,000,000 × 0.11 × 200 = $2,200,000,000 in extra revenue from the premium plan subscriptions. By figuring out the best strategy to deploy, the business maximizes its revenue by making 2 billion extra dollars.

In this Utopian example, we only had two strategies, and besides, we knew their conversion rates. In our case study, we will be facing nine different strategies. Our AI will have no idea of which is the best one, and absolutely no prior information on any of their conversion rates.

We will, however, make the assumption that each of these nine strategies does have a fixed conversion rate. These strategies were carefully and smartly elaborated by the marketing team, and each of them has the same goal: convert the maximum number of clients to the premium plan. However, these nine strategies are all different. They have different forms, different packages, different ads, and different special deals to convince and persuade the clients to subscribe to the premium plan. Of course, the marketing team has no idea of which of these nine strategies will turn out to be the best one. Let's sum up the differences in features of these nine strategies:

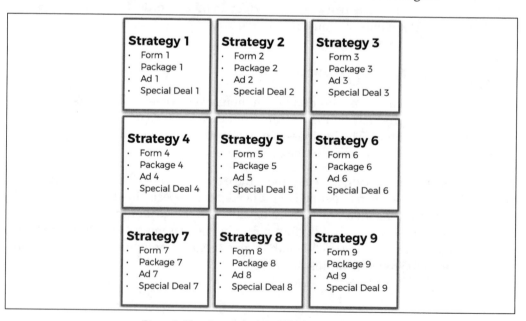

Figure 1: The nine strategies – Which one sells best?

The marketing team wants to figure out which strategy has the highest conversion rate as soon as possible, and by spending the minimum amount. They know that finding and deploying the best strategy can significantly increase the business's revenue. The marketing experts have also chosen not to send an email directly to their 100 million customers, because that would be costly and would risk spamming too many customers. Instead, they will subtly look for that best strategy through online learning. What is online learning? It consists of deploying a different strategy each time a customer browses the e-commerce website.

As the customer navigates the website, they will suddenly get a pop-up ad, suggesting to them that they subscribe to the premium plan. For each customer browsing the website, only one of the nine strategies will be displayed. Then the user will choose, or not, to take action and subscribe to the premium plan. If the customer subscribes, the strategy is a success; otherwise, it is a failure. The more customers we do this with, the more feedback we collect, and the better idea we get of what the best strategy is.

But of course, we will not figure this out manually, visually, or with some simple math. Instead we want to implement the smartest algorithm that will figure out what the best strategy is in the shortest time. That's for the same two reasons: firstly, because deploying each strategy has a cost (for example, coming from the pop-up ad); and secondly, because the company wants to annoy the fewest customers with their ad.

Building the environment inside a simulation

This section is quite special, because there's something crucial to understand which is not obvious at first sight. The reason for this warning is my experience in teaching this subject; many of my students had issues understanding why we have to do a simulation here, for this whole problem.

It was the same for me when I started! If you already understand why we have to make a simulation, that's great—it means you already have online learning under your skin. If not, follow me here and let me explain carefully.

To understand, let's start by explaining what would happen in real life: you would simply display the "call to action" pop-up ad of one of the nine strategies to customers who are navigating the website, and you'd do this one customer at a time. You'd have to do it one customer at a time, customer after customer, because for each customer you need to collect their response: whether or not the customer subscribes to the premium plan. If the customer does, the reward is 1. If not, the reward is 0. It would go like this:

Round 1: We display *Ad 1* of *Strategy 1* to a customer, *Customer 1*, and we check to see if the customer chooses to subscribe. If yes, we get a 1 reward, if no, we get a 0 reward. After collecting our reward, we move on to the next customer (next round).

Round 2: We display *Ad 2* of *Strategy 2* to a new customer, *Customer 2*, and we check to see if the customer chooses to subscribe. If yes, we get a 1 reward, if no, we get a 0 reward. After collecting our reward, we move on to the next customer (next round).

...

Round 9: We display *Ad 9* of *Strategy 9* to a new customer, *Customer 9*, and we check to see if the customer chooses to subscribe. If yes, we get a 1 reward, if no, we get a 0 reward. After collecting our reward, we move on to the next customer (next round).

Round 10: We finally start activating Thompson Sampling! We use the Thompson Sampling AI to tell us which ad has the strongest magic touch to convert the maximum customers to subscribe to the premium plan. We want that extra revenue! The AI (powered by Thompson Sampling) selects one of the 9 ads to display to a new customer, *Customer 10*, and then checks to see if the customer chooses to subscribe. If yes, we get a 1 reward, if no, we get a 0 reward. After collecting our reward, we move on to the next customer (next round).

Round 11: The AI (powered by Thompson Sampling) selects one of the 9 ads to display to a new customer, say *Customer 11*, and then checks to see if the customer chooses to subscribe. If yes, we get a 1 reward, if no, we get a 0 reward. After collecting our reward, we move on to the next customer (next round).

OK, I'll stop! You get the idea. That continues on and on for hundreds of rounds, or at least until the AI has found the best ad — the one with the highest conversion rate.

This is what would happen in real life. We don't need anything else at each round; if you look at the Thompson Sampling algorithm, at each round it only needs the number of times each ad has received a 1 reward in the previous rounds, and the number of times each ad has received a 0 reward in the previous rounds. In conclusion, and this is a very important conclusion: Thompson Sampling absolutely does not need to know the conversion rates of the ads in order to figure out the best ad.

However, in order to simulate this application, we will need to attribute a conversion rate to each of these ads. That's for the simple reason that if we don't do this, we will never be able to verify that Thompson Sampling indeed found the best ad. This is just to check that the AI works!

What we will do is attribute a different conversion rate to each of the nine strategies. The purpose of this simulation will only be to check that the AI manages to catch the best ad, with the highest conversion rate. Let me rephrase this as two essential points:

1. Thompson Sampling at no time needs to know the conversion rates in order to figure out the highest one.

2. The only reason we know the conversion rates in advance is because we are doing a simulation, just to check that Thompson Sampling actually manages to figure out the ad that has the highest conversion rate.

Now we've got that covered, let's finally set these conversion rates. We will assume the nine strategies have the following conversion rates:

Strategy	Conversion Rate
1	0.05
2	0.13
3	0.09
4	0.16
5	0.11
6	0.04
7	0.20
8	0.08
9	0.01

Figure 2: Conversion rates of the 9 strategies

Now, we behind the scenes know in advance which strategy has the highest conversion rate: Strategy number **7**. However, Thompson Sampling doesn't know it. If you pay attention, you can see the fact that at no time does Thompson Sampling use the conversion rates when running its algorithm over the rounds. It only knows the number of successes (subscriptions) and failures (no subscriptions) that have been accumulated over the previous rounds. You can see that most clearly in the code.

Lastly, please make sure to keep in mind that in a real-life situation we would have no idea of what these conversion rates might be. We only know them here for simulation purposes, so that we can check in the end that our AI has managed to figure out the best strategy — which in our simulation here is *Strategy 7*.

The next question is: how exactly are we going to run that simulation?

Running the simulation

First, let's recap the different components of the environment (state, action, and reward):

1. The state is simply a specific customer onto whom we deploy a strategy and show them the ad of that strategy.
2. The action is the strategy selected to be deployed on the customer.
3. The reward is 1 if the customer subscribes to the premium plan, and 0 otherwise.

Then, let's say that this e-commerce business wants to run the experiment of figuring out the best strategy on 10,000 customers. Why the choice of 10,000? Because statistically, this is a large enough sample size to represent the whole base of customers. So, how are we going to simulate the response of 10,000 customers, based on the conversion rates of the ads established before? We don't have a choice other than to take a spreadsheet like Excel, or Google Sheets, and simulate how the 10,000 customers would respond to each of the 9 ads. Here's how we are going to do this; it's a pretty nice trick.

We are going to create a matrix of 10,000 rows and 9 columns. Each row will correspond to a specific customer, and each column will correspond to a specific strategy. To be clear, let's say that:

Row 1 corresponds to *Customer 1*.

Row 2 corresponds to *Customer 2*.

…

Row 10000 corresponds to *Customer 10000*.

Column 1 corresponds to *Strategy 1*.

Column 2 corresponds to *Strategy 2*.

…

Column 9 corresponds to *Strategy 9*.

In the cells of this matrix, we'll place a reward of 1 or 0 depending on whether each of these 10,000 customers would respond positively (subscription) or negatively (no subscription) to each of the 9 strategies. Here's where the "pretty nice trick" comes into play. In order to simulate the response of these 10,000 customers to the 9 ads while considering the conversion rates of these ads, here is what we do:

For each customer (row) and for each strategy (column), we draw a random number between 0 and 1. If this random number is lower than the conversion rate of the strategy, the reward is 1. If this random number is higher than the conversion rate of the strategy, the reward is 0. Why does that work? Because by doing so, we will always have a $p\%$ chance of getting a 1, where p is the conversion rate of the strategy deployed to that customer.

For example, let's take *Strategy 4*, which has a conversion rate of 0.16. For each of the customers, we draw a random number between 0 and 1. That random number has a 16% chance of being between 0 and 0.16, and a (100 – 16) = 84% chance of being between 0.16 and 1. Therefore, since we get a 1 when our random number is between 0 and 0.16, and we get a 0 when our random number is between 0.16 and 1, then that means we have a 16% chance of getting a 1, and an 84% chance of getting a zero.

That simulates exactly the fact that when *Strategy 4* is deployed on a customer, that same customer will have a 16% chance of subscribing to the premium plan; that exactly corresponds to getting a 1 reward.

I hope you like the trick. It's pretty classic, but it's used very often in AI; it's important for you to know about it. We apply that trick to each of the 10,000 x 9 pairs of (customer, strategy) and we get the following matrix (this image only shows the first 10 rows):

	0	1	2	3	4	5	6	7	8
0	0	0	0	0	0	0	0	0	0
1	0	0	0	0	1	0	1	0	0
2	0	0	0	0	0	0	0	0	0
3	0	0	0	1	1	0	0	0	0
4	0	0	0	0	0	0	0	0	0
5	0	0	0	0	0	0	0	0	0
6	0	0	1	0	0	0	1	0	0
7	0	0	0	0	0	0	0	0	0
8	0	0	0	0	0	0	0	0	0
9	0	0	0	0	0	0	0	0	0
10	0	0	0	0	0	0	0	0	0

Figure 3: Simulated matrix of rewards

Let's go through the three first rows in detail:

1. The first customer (row of index **0**) would not subscribe to the premium plan after being approached by any strategy.

2. The second customer (row of index **1**) would subscribe to the premium plan after being approached by *Strategy 5* or *Strategy 7* only.

3. The third customer (row of index **2**) would not subscribe to the premium plan after being approached by any strategy.

We can already see in this preview that our little trick works; the ads with the lowest conversion rates (Strategies 1, 6, and 9) have only 0 rewards for the 11 first customers, while the ads with the highest conversion rates (Strategies 4 and 7) have some 1 rewards already. Note that the indexes here in this Python table start at 0; it's always like that in Python, and unfortunately there is nothing we can do about it. Don't worry, though, you'll get used to it!

If you're a code lover, the code that generated this simulation is presented a little further along in the chapter.

Our next step is to take a step back and recap.

Recap

We're ready to simulate the actions of Thompson Sampling on 10,000 customers successively being approached by one of the 9 strategies, thanks to the preceding matrix, which will exactly simulate the decision of the customer to subscribe or not to the premium plan.

If the cell corresponding to a specific customer and a specific selected strategy has a 1, that simulates a conversion by the customer to the premium plan. If the cell has a 0, that simulates a rejection. Thompson Sampling will collect the feedback of whether or not each of these customers subscribes to the premium plan, one customer after the other. Then, thanks to its powerful algorithm, it will quickly figure out the strategy with the highest conversion rate.

That strategy is the best one to be deployed on millions of customers, maximizing the company's income from this new revenue stream.

AI solution and intuition refresher

Before you enjoy seeing your AI in action, let's refresh our memories and adapt the whole Thompson Sampling AI model to this new problem.

By the way, if you don't like this e-commerce business application, feel totally free to imagine yourself back into the casino, surrounded by nine slot machines having the same conversion rates as the ones given to our strategies. It's exactly the same scenario; the 9 strategies could very well be nine slot machines giving with the same conversion rates either a 1 reward (making you money) or a 0 reward (taking your money). Your goal would be to figure out as quickly as possible which slot machine has the highest chance of giving you the jackpot! It's up to you. Feel free to either go for Vegas or the AI Street, but as far as this chapter is concerned, I'll stick with our e-commerce business.

For starters, let's remind ourselves that each time we show an ad to a new customer, that's considered a new round, n, and we select one of our 9 strategies to attempt a conversion (subscription to the premium plan). The goal is to figure out the best strategy (associated with the ad with the highest conversion rate) in the lowest number of rounds. Here's how Thompson Sampling does that:

AI solution

For each round n over 10,000 rounds, repeat the following three steps:

Step 1: For each strategy i, take a random draw from the following distribution:

$$\theta_i(n) \sim \beta\left(N_i^1(n)+1, N_i^0(n)+1\right)$$

where:

1. $N_i^1(n)$ is the number of times the strategy i has received a 1 reward up to round n.

2. $N_i^0(n)$ is the number of times the strategy i has received a 0 reward up to round n.

Step 2: Select the strategy $s(n)$ that has the highest $\theta_i(n)$:

$$s(n) = \operatorname*{argmax}_{i \in \{1,\dots,9\}}\left(\theta_i(n)\right)$$

Step 3: Update $N_{s(n)}^1(n)$ and $N_{s(n)}^0(n)$ according to the following conditions:

1. If the strategy selected $s(n)$ received a 1 reward:

$$N_{s(n)}^1(n) := N_{s(n)}^1(n) + 1$$

2. If the strategy selected $s(n)$ received a 0 reward:

$$N_{s(n)}^0(n) := N_{s(n)}^0(n) + 1$$

Now we've seen the mathematical steps, let's remind ourselves of the intuition behind them.

Intuition

Each strategy has its own Beta distribution. Over the rounds, the Beta distribution of the strategy with the highest conversion rate will progressively be shifted to the right, and the Beta distributions of the strategies with lower conversion rates will be progressively shifted to the left (*Steps 1 and 3*). Therefore, in *Step 2*, the strategy with the highest conversion rate will be selected more and more often. Here is a graph displaying three Beta distributions of three strategies to help you visualize this:

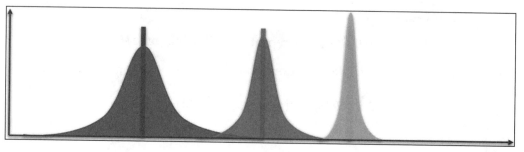

Figure 4: Three Beta distributions

You've taken a step back and you've had a refresher; I think you're ready for the implementation! In the next section, you'll put all that theory into practice—in other words, into code.

Implementation

You'll develop the code as you work along this chapter, but keep in mind that I've provided the whole implementation of Thompson Sampling for this application; you have it available on the GitHub page (`https://github.com/PacktPublishing/AI-Crash-Course`) of this book. If you want to try and run the code, you can do it on Colaboratory, Spyder in Anaconda, or simply your favorite IDE.

Thompson Sampling vs. Random Selection

While implementing Thompson Sampling, you'll also implement the Random Selection algorithm, which will simply select a random strategy at each round. This will be your benchmark to evaluate the performance of your Thompson Sampling model. Of course, Thompson Sampling and the Random Selection algorithm will be competing on the same simulation, that is, on the same environment matrix.

Performance measure

In the end, after the whole simulation is done, you can assess the performance of Thompson Sampling by computing the relative return, defined by the following formula:

$$\text{Relative Return} = \frac{\left(\text{Total Reward of Thompson Sampling}\right) - \left(\text{Total Reward of Random Selection}\right)}{\text{Total Reward of Random Selection}}$$

You'll also have the chance to plot the histogram of selected ads, just to check that the strategy with the highest conversion rate (*Strategy 7*) was the one selected the most.

Let's start coding

First, import the three following required libraries:

1. `numpy`, which you will use to build the environment matrix.
2. `matplotlib.pyplot`, which you will use to plot the histogram.
3. `random`, which you will use to generate the random numbers needed for the simulation.

Here is the extracted code from GitHub:

```
# AI for Sales & Advertizing - Sell like the Wolf of AI Street

# Importing the libraries
import numpy as np
import matplotlib.pyplot as plt
import random
```

Then set the parameters for the number of customers and strategies:

1. N = 10,000 customers.
2. d = 9 strategies.

Code:

```
# Setting the parameters
N = 10000
d = 9
```

Then, create the simulation by building the environment matrix of 10,000 rows corresponding to the customers and 9 columns corresponding to the strategies. At each round, and for each strategy, you draw a random number between 0 and 1. If this random number is lower than the conversion rate of the strategy, the reward will be 1. Otherwise, it will be 0. The environment matrix is named x in the code.

Code:

```
# Building the environment inside a simulation
conversion_rates = [0.05,0.13,0.09,0.16,0.11,0.04,0.20,0.08,0.01]
X = np.array(np.zeros([N,d]))
for i in range(N):
    for j in range(d):
        if np.random.rand() <= conversion_rates[j]:
            X[i,j] = 1
```

Now that the environment is ready, you can start implementing the AI. To do so, the first step is to introduce and initialize the variables you will need for the implementation:

1. `strategies_selected_rs`: A list that will contain the strategies selected over the rounds by the Random Selection algorithm. Initialize it as an empty list.

2. `strategies_selected_ts`: A list that will contain the strategies selected over the rounds by the Thompson Sampling AI model. Initialize it as an empty list.

3. `total_rewards_rs`: The total reward accumulated over the rounds by the Random Selection algorithm. Initialize it as 0.

4. `total_rewards_ts`: The total reward accumulated over the rounds by the Thompson Sampling AI model. Initialize it as 0.

5. `number_of_rewards_1`: A list of 9 elements which will contain for each strategy the number of times it received a 1 reward. Initialize it as a list of 9 zeros.

6. `number_of_rewards_0`: A list of 9 elements which will contain for each strategy the number of times it received a 0 reward. Initialize it as a list of 9 zeros.

Code:

```
# Implementing Random Selection and Thompson Sampling
strategies_selected_rs = []
strategies_selected_ts = []
total_reward_rs = 0
total_reward_ts = 0
numbers_of_rewards_1 = [0] * d
numbers_of_rewards_0 = [0] * d
```

Then you need to begin the `for` loop that will iterate the 10,000 rows (that is, the customers) of this environment matrix. At each round you'll get two separate selections of the deployed strategy; one from the Random Selection algorithm, and one from Thompson Sampling.

Let's start with the Random Selection algorithm, which simply selects a random strategy in each round.

Code:

```
for n in range(0, N):
    # Random Selection
    strategy_rs = random.randrange(d)
    strategies_selected_rs.append(strategy_rs)
    reward_rs = X[n, strategy_rs]
    total_reward_rs = total_reward_rs + reward_rs
```

Next, you need to implement Thompson Sampling following exactly *Steps 1, 2*, and *3* provided previously. I recommend looking at these steps again before coding the next part, and try to code by yourself before seeing my solution. That's the best way you can progress; practice makes perfect. You have all the elements required to code this; you even have similar code in *Chapter 5, Your First AI Model – Beware the Bandits!*. Good luck! Here is the solution.

You should implement Thompson Sampling step by step, starting with the first step. Let's remind ourselves of it:

Step 1: For each strategy i, take a random draw from the following distribution:

$$\theta_i(n) \sim \beta\left(N_i^1(n)+1, N_i^0(n)+1\right)$$

where:

1. $N_i^1(n)$ is the number of times the strategy *i* has received a 1 reward up to round *n*

2. $N_i^0(n)$ is the number of times the strategy *i* has received a 0 reward up to round *n*

Let's see how *Step 1* is implemented.

Code a second `for` loop that iterates the 9 strategies, because you have to take a random draw from the Beta distribution of each of the 9 strategies.

The random draws from the Beta distributions are generated by the `betavariate()` function taken from the `random` library, which you imported at the beginning.

Code:

```
# Thompson Sampling
strategy_ts = 0
max_random = 0
for i in range(0, d):
    random_beta = random.betavariate(numbers_of_rewards_1[i] + 1,
numbers_of_rewards_0[i] + 1)
```

Now implement *Step 2*, that is:

Step 2: Select the strategy $s(n)$ that has the highest $\theta_i(n)$:

$$s(n) = \underset{i \in \{1,\dots,9\}}{\mathrm{argmax}} \left(\theta_i(n) \right)$$

To implement *Step 2*, you stay in the second `for` loop which iterates the 9 strategies, and use a simple trick with an `if` condition that will figure out the highest $\theta_i(n)$.

The trick is the following: while iterating the strategies, if you find a random draw (`random_beta`) that is higher than the maximum of the random draws obtained so far (`max_random`), then that maximum becomes equal to that higher random draw.

Code:

```
# Thompson Sampling
strategy_ts = 0
max_random = 0
for i in range(0, d):
    random_beta = random.betavariate(numbers_of_rewards_1[i] + 1,
numbers_of_rewards_0[i] + 1)
```

```
    if random_beta > max_random:
        max_random = random_beta
        strategy_ts = i
reward_ts = X[n, strategy_ts]
```

And finally, let's implement *Step 3*, the easiest one:

Step 3: Update $N^1_{s(n)}(n)$ and $N^0_{s(n)}(n)$ according to the following conditions:

1. If the strategy selected $s(n)$ received a 1 reward:

$$N^1_{s(n)}(n) := N^1_{s(n)}(n) + 1$$

2. If the strategy selected $s(n)$ received a 0 reward:

$$N^0_{s(n)}(n) := N^0_{s(n)}(n) + 1$$

Implement that simply with the exact same two `if` conditions, translated into code.

Code:

```
# Thompson Sampling
strategy_ts = 0
max_random = 0
for i in range(0, d):
    random_beta = random.betavariate(numbers_of_rewards_1[i] + 1,
numbers_of_rewards_0[i] + 1)
    if random_beta > max_random:
        max_random = random_beta
        strategy_ts = i
reward_ts = X[n, strategy_ts]
if reward_ts == 1:
    numbers_of_rewards_1[strategy_ts] = numbers_of_
rewards_1[strategy_ts] + 1
else:
    numbers_of_rewards_0[strategy_ts] = numbers_of_
rewards_0[strategy_ts] + 1
```

Next, don't forget to add the strategy selected in *Step 2* to our list of strategies (`strategies_selected_ts`), and also to compute the total reward accumulated over the rounds by Thompson Sampling (`total_reward_ts`).

Code:

```
# Thompson Sampling
strategy_ts = 0
max_random = 0
for i in range(0, d):
    random_beta = random.betavariate(numbers_of_rewards_1[i] + 1,
numbers_of_rewards_0[i] + 1)
    if random_beta > max_random:
        max_random = random_beta
        strategy_ts = i
reward_ts = X[n, strategy_ts]
if reward_ts == 1:
    numbers_of_rewards_1[strategy_ts] = numbers_of_
rewards_1[strategy_ts] + 1
else:
    numbers_of_rewards_0[strategy_ts] = numbers_of_
rewards_0[strategy_ts] + 1
strategies_selected_ts.append(strategy_ts)
total_reward_ts = total_reward_ts + reward_ts
```

Then compute the final score, which is the relative return of Thompson Sampling with respect to our benchmark, which is Random Selection:

Code:

```
# Computing the Relative Return
relative_return = (total_reward_ts - total_reward_rs) / total_reward_
rs * 100
print("Relative Return: {:.0f} %".format(relative_return))
```

The final result

By executing this code, I obtained a final relative return of 91%. In other words, Thompson Sampling almost doubled the performance of my Random Selection benchmark. Not too bad!

Finally, plot a histogram of the selected strategies to check that *Strategy 7* (at index 6) was the one most selected, since it is the one with the highest conversion rate. To do this, use the hist() function from the matplotlib library.

Code:

```
# Plotting the Histogram of Selections
plt.hist(strategies_selected_ts)
plt.title('Histogram of Selections')
```

```
plt.xlabel('Strategy')
plt.ylabel('Number of times the strategy was selected')
plt.show()
```

This is the most exciting time—the code is complete (congrats by the way), and you can enjoy the results. Having the final relative return is nice, but finishing with a clean visualization plot is even better. And that's what you get by executing the final code:

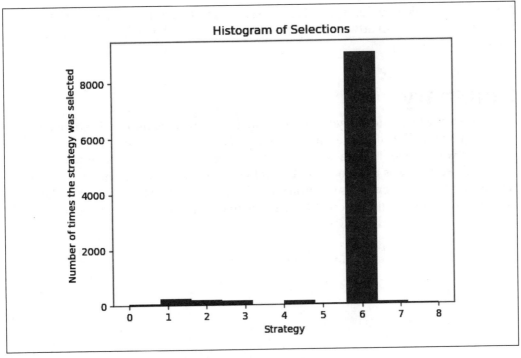

Figure 5: Histogram of Selections

You can see that the strategy at index **6**, *Strategy 7*, was by far selected the most. Thompson Sampling was quickly able to identify it as the best strategy. In fact, if you re-run the same code but with only 1,000 customers, you'll see that Thompson Sampling is still able to identify *Strategy 7* as the best one.

Thompson Sampling did an amazing job for this e-commerce business. Not only was it able to identify the best strategy in a small number of rounds—that means fewer customers, which saves on advertising and operating costs—but also it was able to clearly figure out the strategy with the highest conversion rate.

If this e-commerce business has, for example, 50 million customers, and if the premium plan has a price of $200 per year, then deploying this best strategy with a conversion rate of 20 % would lead to generate an extra revenue of 50,000,000 × 0.2 × $200 = $2 billion!

In other words, Thompson Sampling clearly and quickly smashed the sales and advertising for this e-commerce business, so much so that we really can call it the wolf of AI Street.

Now, take a break, you deserve it. Get refreshed, and as soon as you are recharged and all set for a new AI adventure, I'll be here ready as well to start the next chapter. See you back soon!

Summary

In this first practical tutorial, you implemented Thompson Sampling to solve the multi-armed bandit problem as applied to an advertising campaign. Thompson Sampling was able to find the best business strategy quickly, something which Random Selection was unable to do. In total you generated 91% of relative return, which, after making some assumptions, would generate an extra 2 billion dollars in revenue. You did that in just one file in less than 60 lines of code. Quite astounding, right?

7
Welcome to Q-Learning

Ladies and gentlemen, things are about to get even more interesting than before. The next model we are about to tackle is at the heart of many AIs built today; robots, autonomous vehicles, and even AI players of video games. They all use Q-learning at the core of their model. Some of them even combine Q-learning with deep learning, making a highly advanced version of Q-learning called deep Q-learning, which we will cover in *Chapter 9, Going Pro with Artificial Brains – Deep Q-Learning*.

All of the AI fundamentals still apply to Q-learning, as follows:

1. Q-learning is a Reinforcement Learning model.
2. Q-learning works on the inputs (states) and outputs (actions) principle.
3. Q-learning works on a predefined environment, including the states (the inputs), the actions (the outputs), and the rewards.
4. Q-learning is modeled by a Markov decision process.
5. Q-learning uses a training mode, during which the parameters that are learned are called the Q-values, and an inference mode.

Now we can add two more fundamentals, this time specific to Q-learning:

1. There are a finite number of states (there is not an infinity of possible inputs).
2. There are a finite number of actions (only a certain number of actions can be performed).

That's all! There are no more fundamentals to keep in mind; now we can really dig into Q-learning, which you'll see is not that hard and really quite intuitive.

To explain Q-learning, we'll use an example so that you won't get lost inside pure theory, and so that you can visualize what's happening. On that note: welcome to the Maze.

The Maze

You are going to learn how Q-learning works inside a maze. Let's draw our maze right away; here it is:

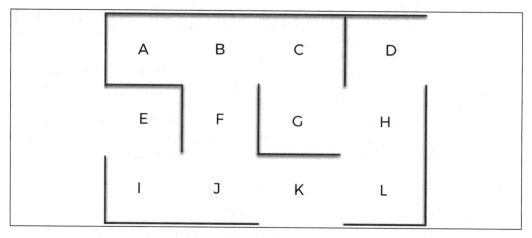

Figure 1: The Maze

I know, it's the simplest maze you have ever seen. That's important for the sake of simplicity, so that you can mostly focus on how the AI works its magic. Imagine if you got lost in this chapter because of the maze and not because of the AI formulas! The important thing is that you have a clear maze, and you can visualize how the AI might manage to find its way from the beginning to the end.

Speaking of the beginning and the end, imagine a little robot inside this maze, starting at point **E** (Entrance). Its goal is to find the quickest way to point **G** (Goal). We humans can figure that out in no time, but that's only because our maze is so simple. What you are going to build is an AI that can go from a starting point to an ending point, regardless of how complex the maze is. Let's get started!

Beginnings

Here is a question for you: what do you think is going to be the very first step?

I'll give you three possible answers:

1. We start writing some math equations.

2. We build the environment.

3. We try to make it work with Thompson Sampling (the AI model of the previous chapter).

The correct answer is…

2. We build the environment.

That was easy, but I wanted to highlight that in a question to make sure you keep in mind that this must always be the first step when building an AI. After clearly understanding the problem, the first step of building your AI solution is always to set up the environment.

That begs a further question:

What steps, exactly, are you going to take when building that environment?

Try to remember the answer—you've already learned this—and then read on for a recap.

1. Firstly, you'll define the states (the inputs of your AI).

2. Secondly, you'll define the actions that can be performed (the outputs of your AI).

3. Thirdly, you'll define the rewards. Remember, the reward is what the AI gets after performing an action in a certain state.

Now we've secured the basics, so you can tackle that first step of defining the environment.

Building the environment

To build the environment, we need to define the states, the actions, and the rewards.

The states

Let's begin with the states. What do you think are going to be the states for this problem? Remember, the states are the inputs of your AI. And they should contain enough information for the AI to be able to take an action that will lead it to its final goal (reaching point E).

In this model, we don't have too much of a choice. The state, at a specific time or specific iteration, is simply going to be the position of the AI at that time. In other words, it is going to be the letter of the location, from **A** to **L**, where the AI is in at a specific time.

As you might guess, the next step after building the environment will be writing the mathematical equations at the heart of the AI, and to help you with that, it makes it much easier to encode the states into unique integers instead of keeping them as letters. That's exactly what we are going to do, with the following mapping:

Location	State
A	0
B	1
C	2
D	3
E	4
F	5
G	6
H	7
I	8
J	9
K	10
L	11

Figure 2: Location to state mapping

Notice that we abide by the first specific fundamental of Q-learning, that is: **there are a finite number of states**.

Let's move on to the actions.

The actions

The actions are simply going to be the next moves the AI can make to go from one location to the next. For example, let's say the AI is in location **J**; the possible actions that the AI can perform are to go to **I**, to **F**, or to **K**. Again, since you'll be working with math equations, you can encode these actions with the same indexes as for the states.

Following the example where the AI is in location **J** at a specific time, the possible actions that the AI can perform are **5**, **8**, and **10**, according to our previous mapping above: the index **5** corresponds to **F**, the index **8** corresponds to **I**, and the index **10** corresponds to **K**.

Hence, the possible actions are simply the indexes of the different locations that can be reached:

Possible actions = {0,1,2,3,4,5,6,7,8,9,10,11}

Notice that again, we abide by the second specific fundamental of Q-learning, that is: **there are a finite number of actions**.

Now obviously, when in a specific location, there are some actions that the AI cannot perform. Taking the same previous example, if the AI is in location **J**, it can perform the actions **5**, **8**, and **10**, but it cannot perform the other actions. You can make sure to specify that by attributing a 0 reward to the actions it cannot perform, and a 1 reward to the actions it can perform. That brings us to the rewards.

The rewards

You're almost done with your environment — last, but not least, you have to define a system of rewards. More specifically, you have to define a reward function R that takes as input a state s and an action a, and returns a numerical reward r that the AI will get by performing the action a in the state s:

$$R: (s, a) \mapsto r \in \mathbf{R}$$

So, how can you build such a function for our case study? Here, it is simple. Since there are a discrete and finite number of states (the indexes from 0 to 11), as well as a discrete and finite number of actions (same indexes from 0 to 11), the best way to build your reward function R is to simply make a matrix.

Your reward function will be a matrix of exactly 12 rows and 12 columns, where the rows correspond to the states, and the columns correspond to the actions. That way, in your function R: (s, a) $\mapsto r \in \mathbf{R}$, s will be the row index of the matrix, a will be the column index of the matrix, and r will be the cell of index (s, a) in the matrix.

To build this reward matrix, what you first have to do is attribute, for each of the 12 locations, a 0 reward to the actions that the robot cannot perform, and a 1 reward to the actions the robot can perform. By doing that for each of the 12 locations, you will end up with a matrix of rewards. Let's build it step by step, starting with the first location: location **A**.

When in location **A**, the robot can only go to location **B**. Therefore, since location **A** has index 0 (first row of the matrix) and location **B** has index 1 (second column of the matrix), the first row of the matrix of rewards will get a 1 on the second column, and a 0 on all the other columns, like so:

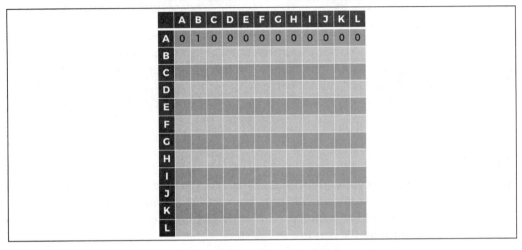

Figure 3: Rewards matrix – Step 1

Let's move on to location **B**. When in location **B**, the robot can only go to three different locations: **A**, **C**, and **F**. Since **B** has index 1 (second row), and **A**, **C**, and **F** have respective indexes 0, 2, and 5 (1st, 3rd, and 6th column), then the second row of the matrix of rewards will get a 1 on the 1st, 3rd, and 6th columns, and 0 on all the other columns:

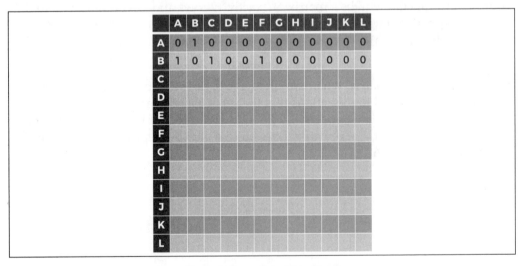

Figure 4: Rewards matrix – Step 2

C (of index 2) is only connected to **B** and **G** (of indexes 1 and 6) so the third row of the matrix of rewards is:

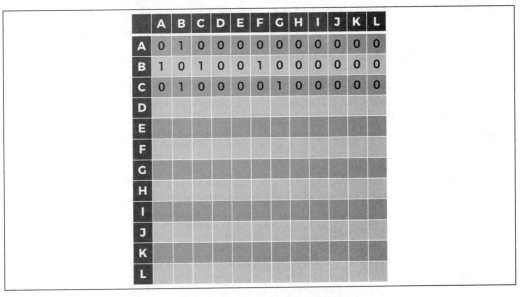

Figure 5: Rewards matrix – Step 3

By doing the same for all the other locations, you eventually get your final matrix of rewards:

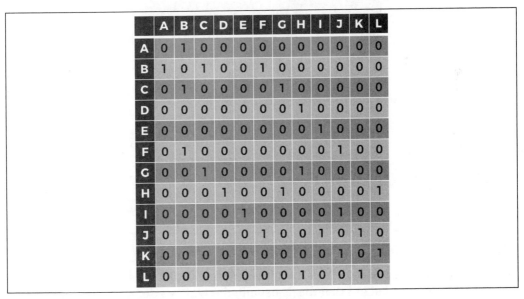

Figure 6: Rewards matrix - Step 4

And that's how you initialize the matrix of rewards.

But wait—you're not actually finished. There is one final thing you need to do. It's a step that's crucial to understand. In fact, let me ask you another question, the ultimate one, which will check if your intuition is already shaping up:

How can you let the AI know that it has to go to that top priority location G?

It's easy—you do it simply by playing with the rewards. You must keep in mind that with Reinforcement Learning, everything works from the rewards. If you attribute a high reward to location **G**, for example 1000, then the AI will automatically try to go and catch that high reward, simply because it is larger than the rewards of the other locations.

In short, and it's a fundamental point to understand and remember in Reinforcement Learning in general, **the AI is always looking for the highest reward**. That's why the trick to reach location **G** is simply to attribute it a higher reward than the other locations.

For now, manually put a high reward (1000) inside the cell corresponding to location **G**, because it is the goal location where we want our AI to go. Since location **G** has an index of 6, we put a 1000 reward on the cell of row 6 and column 6. Accordingly, our matrix of rewards becomes:

	A	B	C	D	E	F	G	H	I	J	K	L
A	0	1	0	0	0	0	0	0	0	0	0	0
B	1	0	1	0	0	1	0	0	0	0	0	0
C	0	1	0	0	0	0	1	0	0	0	0	0
D	0	0	0	0	0	0	0	1	0	0	0	0
E	0	0	0	0	0	0	0	0	1	0	0	0
F	0	1	0	0	0	0	0	0	0	1	0	0
G	0	0	1	0	0	0	1000	1	0	0	0	0
H	0	0	0	1	0	0	1	0	0	0	0	1
I	0	0	0	0	1	0	0	0	0	1	0	0
J	0	0	0	0	0	1	0	0	1	0	1	0
K	0	0	0	0	0	0	0	0	0	1	0	1
L	0	0	0	0	0	0	0	1	0	0	1	0

Figure 7: Rewards matrix - Step 5

You have defined the rewards! You did it by simply building this matrix of rewards. It is important to understand that this is usually the way we define the system of rewards when doing Q-learning.

In *Chapter 9, Going Pro with Artificial Brains – Deep Q-Learning,* which is about deep Q-learning, you will see that we will proceed very differently and build the environment much more easily. In fact, deep Q-learning is the advanced version of Q-learning that is widely used today in AI, far more than the simple Q-learning model itself. But you have to tackle Q-learning first, in depth, in order to be ready for deep Q-learning.

Since you've defined the states, the actions, and the rewards, you have finished building the environment. This means you are ready to tackle the next step, where you will build the AI itself that will do its magic inside this environment that you've just defined.

Building the AI

Now that you have built an environment in which you clearly defined the goal with a relevant system of rewards, it's time to build the AI. I hope you're ready for a little math.

I'll break down this second step into several sub-steps, leading you to the final Q-learning model. To that end, we'll cover three important concepts at the heart of Q-learning, in the following order:

1. The Q-value
2. The temporal difference
3. The Bellman equation

Let's get started by learning about the Q-value.

The Q-value

Before you start getting into the details of Q-learning, I need to explain the concept of the Q-value. Here's how it works:

To each couple of state and action (s, a), we are going to associate a numeric value $Q(s, a)$:

$$Q:\left(s \in S, a \in A\right) \mapsto Q\left(s,a\right) \in \mathbb{R}$$

We will say that $Q(s, a)$ is "the Q-value of the action a performed in the state s."

Now I know the sort of questions you might be asking in your head: What does this Q-value mean? What does it represent? How do I even compute it? These were some of the questions I had in my mind when I first learned Q-learning.

In order to answer these questions, I need to introduce the temporal difference.

The temporal difference

This is where the math really comes in. Let's say we are in a specific state s_t, at a specific time t. Let's just perform an action randomly, any of them. That brings us to the next state s_{t+1} and we get the reward $R(s_t, a_t)$.

The temporal difference at time t, denoted by $TD_t(s_t, a_t)$, is the difference between:

1. $R(s_t, a_t) + \gamma \max_a (Q(s_{t+1}, a))$, that is, the reward $R(s_t, a_t)$ obtained by performing the action a_t in the state s_t, plus the Q-value of the best action performed in the future state s_{t+1}, discounted by a factor $\gamma \in [0,1]$, called the discount factor

2. and $Q(s_t, a_t)$, that is, the Q-value of the action a_t performed in the state s_t.

This leads to:

$$TD_t(s_t, a_t) = R(s_t, a_t) + \gamma \max_a (Q(s_{t+1}, a)) - Q(s_t, a_t)$$

You might think that's great, that you understand all the terms, but you're probably also thinking "But what does that all mean?" Don't worry—that's exactly what I was thinking when I was learning this.

I'm going to explain while at the same time improving your AI intuition. The first thing to understand is that the temporal difference represents how well the AI is learning. Here's how it works exactly, with respect to the training process (during which the Q-values are learned):

1. At the beginning of the training, the Q-values are set to zero. Since the AI is looking to get the good rewards (here 1 or 1000), it is looking for the high temporal differences (see the formula of TD). Accordingly, if in the first iterations, $TD_t(s_t, a_t)$ is high, the AI gets a "pleasant surprise" because that means the AI was able to find a good reward. On the other hand, if $TD_t(s_t, a_t)$ is small, the AI gets a "frustration."

2. When the AI gets a great reward, the specific Q-value of the (state, action) that led to that great reward increases, so the AI can remember how it got to that high reward (you'll see exactly how it increases in the next section). For example, let's say that it was the action a_t performed in the state s_t that led to that high reward $R(s_t, a_t)$. That would mean the Q-value $Q(s_t, a_t)$ increases automatically (remember, you'll see how in the next section). Those increased Q-values are important information, because they indicate to the AI which transitions lead to the good rewards.

3. The next step of the AI is not only to look for the great rewards, but also to look at the same time for the high Q-values. Why? Because the high Q-values are the ones that lead to the great reward. In fact, the high Q-values are the ones that lead to higher Q-values, themselves leading to even higher Q-values, themselves leading eventually to the highest reward (1000). That's the role of $\gamma \max_a \left(Q\left(s_{t+1}, a\right) \right)$ in the temporal difference formula. Everything will become crystal clear when you put this into practice. The AI looks for the high Q-values, and as soon as it finds them, the Q-values of the (state, action) that led to these high Q-values will increase again, since they indicate the right path towards the goal.

4. At some point, the AI will know all the transitions that lead to the good rewards and high Q-values. Since the Q-values of these transitions have already been increased over time, the temporal differences decrease in the end. In fact, the closer we get to the final goal, the smaller the temporal differences become.

In conclusion, the temporal difference is like a temporary intrinsic reward, of which the AI will try to find the large values at the beginning of the training. Eventually, the AI will minimize this reward as it gets to the end of the training—that is, as it gets closer to the final goal.

That's exactly the intuition of the temporal difference you must have in mind, because it will really help you understand the magic of Q-learning. Speaking of that magic, we are about to reveal the last piece of the puzzle.

Now you understand that the AI will iterate some updates of the Q-values towards the high temporal differences, which are ultimately decreased. But how does it do that? There is a specific answer to that question—the Bellman equation, the most famous equation in Reinforcement Learning.

The Bellman equation

In order to perform better and better actions that will lead the AI to reach its goal, you have to increase the Q-values of actions when you find high temporal differences. Only one question remains: How will the AI update these Q-values? Richard Bellman, a pioneer of Reinforcement Learning, created the answer. At each iteration, you update the Q-values from time t-1 (previous iteration) to t (current iteration) through the following equation, called the Bellman equation:

$$Q_t(s_t, a_t) = Q_{t-1}(s_t, a_t) + \alpha TD_t(s_t, a_t)$$

where $\alpha \in \mathbb{R}$ is the learning rate, which dictates how fast the learning of the Q-values goes. Its value is usually between 0 and 1, for example, 0.75. The lower the value of α, the smaller the updates of the Q-values, and the longer the Q-learning will take. The higher its value, the bigger the updates of the Q-values and the faster the Q-learning will be. As you can clearly see in this equation, when the temporal difference $TD_t\left(s_t, a_t\right)$ is high, the Q-value $Q_t\left(s_t, a_t\right)$ increases.

Reinforcement intuition

Now you have all the elements of Q-learning—congratulations, by the way— let's connect the dots between all these elements to reinforce your AI intuition.

The Q-values measure the accumulation of "good surprise" or "frustration" associated with the couple of action and state $\left(s_t, a_t\right)$.

In the "good surprise" case of a high temporal difference, the AI is reinforced, and in the "frustration" case of a low temporal difference, the AI is weakened.

We want to learn the Q-values that will give the AI the maximum "good surprise," and that's exactly what the Bellman equation does by updating the Q-values at each iteration.

You've learned quite a lot of new information, and even though you've finished with an intuition section that connects the dots, that's not enough to get a really solid grasp of Q-learning. The next step is to take a step back, and the best way to do that is to go through the whole Q-learning process from start to finish so that it becomes crystal clear in your head.

The whole Q-learning process

Let's summarize the different steps of the whole Q-learning process. To be clear, the only purpose of this process is to update the Q-values over a certain number of iterations until they are no longer updated (we refer to that point as convergence).

The number of iterations depends on the complexity of the problem. For our problem, 1,000 will be enough, but for more complex problems you might want to consider higher numbers such as 10,000. In short, the Q-learning process is the part where we train our AI, and it's called Q-learning because it's the process during which the Q-values are learned. Then I'll explain what happens for the inference part (pure predictions), which comes, as always, after the training. The full Q-learning process starts with training mode.

Training mode

Initialization (First iteration):

For all couples of states *s* and actions *a*, the Q-values are initialized to 0.

Next iterations:

At each iteration $t \geq 1$, you repeat for a certain number of times (chosen by you the developer) the following steps:

1. You select a random state s_t from the possible states.
2. From that state, you perform a random action a_t that can lead to a next possible state, that is, such that $R(s_t, a_t) > 0$.
3. You reach the next state s_{t+1} and you get the reward $R(s_t, a_t)$.
4. You compute the temporal difference $TD_t(s_t, a_t)$:

$$TD_t(s_t, a_t) = R(s_t, a_t) + \gamma \max_a \left(Q(s_{t+1}, a) \right) - Q(s_t, a_t)$$

5. You update the Q-value by applying the Bellman equation:

$$Q_t(s_t, a_t) = Q_{t-1}(s_t, a_t) + \alpha TD_t(s_t, a_t)$$

At the end of this process, you have obtained Q-values that no longer update. That means only one thing; you are ready to hack the maze by going into inference mode.

Inference mode

The training is complete, and now begins the inference. To remind you, the inference part is when you have a fully trained model with which you are going to make predictions. In our maze, the predictions that you are going to make are the actions to perform to take you from start (Location **E**) to finish (Location **G**). So, the question is:

How are you going to use the learned Q-values to perform the actions?

Good news; for Q-learning this is very simple. When in a certain state s_t, you simply perform the action a_t that has the highest Q-value for that state s_t:

$$a_t = \arg\max_a \left(Q(s_t, a) \right)$$

That's all—by doing this at each location (each state), you get to your final destination through the shortest route. We'll implement this and see the result in the practical activities or the next chapter.

Summary

In this chapter we studied the Q-learning model, which is only applied to environments that have a finite number of input states and a finite number of possible actions to perform.

When performing Q-learning, the AI learns Q-values through an iterative process, so that the higher the Q-value of a (state, action) pair, the closer the AI gets to the top reward.

At each iteration the Q-values are updated through the Bellman equation, which simply consists of adding the temporal difference, discounted by a learning rate factor. We will get to work on a full practical Q-learning activity in the next chapter, applied to a real-world business problem.

8
AI for Logistics – Robots in a Warehouse

It's time for the next step on our AI journey. I told you at the beginning of this book that AI has tremendous value to bring to transport and logistics, with self-driving delivery vehicles that speed up logistical processes. They're a huge boost to the economy through the e-commerce industry.

In this new chapter, we'll build an AI for just that kind of application. The model we'll use for this will, of course, be Q-learning (we're saving deep Q-learning for the self-driving car). Q-learning is a simple, but powerful, AI model that can optimize the flows of movement in a warehouse, which is the real-world problem you'll solve here. In order to facilitate this journey, you'll work on an environment you're already familiar with: the maze we saw in the previous chapter.

The difference is that, this time, the maze will actually be the warehouse of a business. It could be any business: an e-commerce business, a retail business, or any business that sells products to customers and that has a warehouse to store large amounts of products to be sold.

Let's have a look again at this maze, now a warehouse:

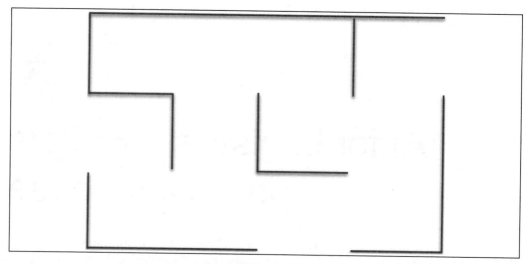

Figure 1: The warehouse

Inside this warehouse, the products are stored in 12 different locations, labeled by the following letters from **A** to **L**:

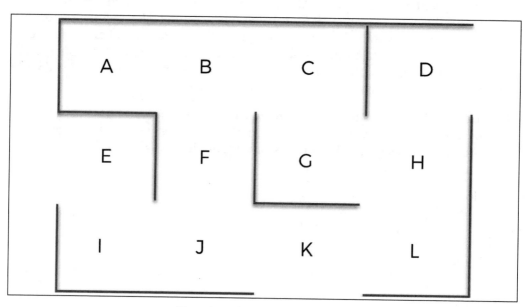

Figure 2: Locations in the warehouse

When orders are placed by customers, a robot moves around the warehouse to collect the products for delivery. That will be your AI! Here's what it looks like:

Figure 3: Warehouse robot

The 12 locations are all connected to a computer system, which ranks in real time the product collection priorities for these 12 locations. As an example, let's say that at a specific time, *t*, it returns the following ranking:

Priority Rank	Location
1	G
2	K
3	L
4	J
5	A
6	I
7	H
8	C
9	B
10	D
11	F
12	E

Figure 4: Top priority locations

Location **G** has priority 1, which means it's the top priority, as it contains a product that must be collected and delivered immediately. Our robot must move to location **G** by the shortest route depending on where it is. Our goal is to actually build an AI that will return that shortest route, wherever the robot is.

But we could do even better. Here, locations **K** and **L** are in the top 3 priorities. Hence, it would be great to implement an option for our robot to go via some intermediary locations before reaching its final top priority location.

The way the system computes the priorities of the locations is outside the scope of this case study. The reason for this is that there can be many ways, from simple rules or algorithms, to deterministic computations, to machine learning, to compute these priorities. But most of these ways would not be AI as we know it today. What we really want to focus on in this exercise is the core AI, encompassing Reinforcement Learning and Q-learning. We can just say for the purposes of this example that location **G** is the top priority because one of the most loyal platinum-level customers of the company placed an urgent order of a product stored in location **G**, which therefore must be delivered as soon as possible.

In conclusion, our mission is to build an AI that will always take the shortest route to the top priority location, whatever the location it starts from, and have the option to go by an intermediary location which is in the top three priorities.

Building the environment

When building an AI, the first thing we always have to do is define the environment. Defining an environment always requires the following three elements:

- Defining the states
- Defining the actions
- Defining the rewards

These three elements have already been defined in the previous chapter on Q-learning, but let's quickly remind ourselves what they are.

The states

The state, at a specific time t, is the location where the robot is at that time t. However, remember, you have to encode the location names so that our AI can do the math.

At the risk of disappointing you, given all the crazy hype about AI, let's remain realistic and understand that Q-learning is nothing more than a bunch of math equations; just like any other AI model. Let's make the encoding integers start at 0, simply because indexes in Python start at 0:

Location	State
A	0
B	1
C	2
D	3
E	4
F	5
G	6
H	7
I	8
J	9
K	10
L	11

Figure 5: Location to state mapping

The actions

The actions are the next possible destinations to which the robot can go. You can encode these destinations with the same indexes as the states. Hence, the total list of actions that the AI can perform is the following:

```
actions = [0,1,2,3,4,5,6,7,8,9,10,11]
```

The rewards

Remember, when in a specific location, there are some actions that the robot cannot perform. For example, if the robot is in location **J**, it can perform the actions 5, 8, and 10, but it cannot perform the other actions. You can specify that by attributing a reward of 0 to the actions it cannot perform, and a reward of 1 to the actions it can perform.

That brings you to building the following matrix of rewards:

	A	B	C	D	E	F	G	H	I	J	K	L
A	0	1	0	0	0	0	0	0	0	0	0	0
B	1	0	1	0	0	1	0	0	0	0	0	0
C	0	1	0	0	0	0	1	0	0	0	0	0
D	0	0	0	0	0	0	0	1	0	0	0	0
E	0	0	0	0	0	0	0	0	1	0	0	0
F	0	1	0	0	0	0	0	0	0	1	0	0
G	0	0	1	0	0	0	0	1	0	0	0	0
H	0	0	0	1	0	0	1	0	0	0	0	1
I	0	0	0	0	1	0	0	0	0	1	0	0
J	0	0	0	0	0	1	0	0	1	0	1	0
K	0	0	0	0	0	0	0	0	0	1	0	1
L	0	0	0	0	0	0	0	1	0	0	1	0

Figure 6: Rewards matrix

AI solution refresher

It never hurts to get a little refresher of a model before implementing it! Let's remind ourselves of the steps of the Q-learning process; this time, adapting it to your new problem. Let's welcome Q-learning back on stage:

Initialization (first iteration)

For all pairs of states *s* and actions *a*, the Q-values are initialized to 0:

For all states $s = 0,\ldots,11$ and actions $a = 0,\ldots,11: Q(s,a) = 0$

Next iterations

At each iteration $t \geq 1$, the AI will repeat the following steps:

1. It selects a random state s_t from the possible states:

$$s_t = \text{random}(0,1,2,3,4,5,6,7,8,9,10,11)$$

2. It performs a random action a_t that can lead to a next possible state, that is, such that $R(s_t, a_t) > 0$:

$$a_t = \text{random}(0,1,2,3,4,5,6,7,8,9,10,11)\,\text{s.t.}\,R(s_t, a_t) > 0$$

3. It reaches the next state s_{t+1} and gets the reward $R(s_t, a_t)$.

4. It computes the temporal difference $TD_t(s_t, a_t)$:

$$TD_t(s_t, a_t) = R(s_t, a_t) + \gamma \max_a \left(Q(s_{t+1}, a)\right) - Q(s_t, a_t)$$

5. It updates the Q-value by applying the Bellman equation:

$$Q_t(s_t, a_t) = Q_{t-1}(s_t, a_t) + \alpha TD_t(s_t, a_t)$$

We repeat these steps over 1,000 iterations. Why 1,000? The choice of 1,000 comes from my experimentation with this particular environment. I chose a number that's large enough for the Q-values to converge over the training. 100 wasn't large enough, but 1,000 was. Usually, you can just pick a very large number, for example, 5,000, and you will get convergence (that is, the Q-values will no longer update). However, that depends on the complexity of the problem. If you are dealing with a much more complex environment, for example, if you had hundreds of locations in the warehouse, you'd need a much higher number of training iterations.

That's the whole process. Now, you're going to implement it in Python from scratch!

Are you ready? Let's do this.

Implementation

Alright, let's smash this. But first, try to smash this yourself without me. Of course, this is a journey we'll take together, but I really don't mind if you take some steps ahead of me. The faster you become independent in AI, the sooner you'll do wonders with it. Try to implement the Q-learning process mentioned previously, exactly as it is. It's okay if you don't implement everything; what matters is that you try.

That's enough coaching; no matter how successful you were, let's go through the solution.

First, start by importing the libraries that you'll use in this implementation. There's only one needed this time: the numpy library, which offers a practical way of working with arrays and mathematical operations. Give it the shortcut np.

```
# AI for Logistics - Robots in a warehouse

# Importing the libraries
import numpy as np
```

Then, set the parameters of your model. These include the discount factor γ and the learning rate α, which are the only parameters of the Q-learning model. Give them the values of 0.75 and 0.9 respectively, which I've arbitrarily picked but are usually a good choice. These are decent values to start with if you don't know what to use. However, you'll get the same result with similar values.

```
# Setting the parameters gamma and alpha for the Q-Learning
gamma = 0.75
alpha = 0.9
```

The two previous code sections were simply the introductory sections, before you really start to build your AI model. The next step is to start the first part of our implementation.

Try to remember what you have to do now, as a first general step of building an AI.

You build the environment!

I just wanted to highlight that, once again; it's really compulsory. The environment will be the first part of your code:

Part 1 – Building the environment

Let's look at the whole structure of this implementation so that you can take a step back already. Your code will be structured in three parts:

- **Part 1** – Building the environment
- **Part 2** – Building the AI solution with Q-learning (training)
- **Part 3** – Going into production (inference)

Let's start with part 1. For that, you define the states, the actions, and the rewards. Begin by defining the states, with a Python dictionary mapping the location's names (in letters from A to L) into the states (in indexes from 0 to 11). Call this dictionary `location_to_state`:

```
# PART 1 - BUILDING THE ENVIRONMENT

# Defining the states
location_to_state = {'A': 0,
                     'B': 1,
                     'C': 2,
                     'D': 3,
                     'E': 4,
                     'F': 5,
                     'G': 6,
                     'H': 7,
                     'I': 8,
                     'J': 9,
                     'K': 10,
                     'L': 11}
```

Then, define the actions with a simple list of indexes from 0 to 11. Remember that each action index corresponds to the next location where that action leads to:

```
# Defining the actions
actions = [0,1,2,3,4,5,6,7,8,9,10,11]
```

Finally, define the rewards, by creating a matrix of rewards where the rows correspond to the current states s_t, the columns correspond to the actions a_t leading to the next state s_{t+1}, and the cells contain the rewards $R(s_t, a_t)$. If a cell (s_t, a_t) contains a 1, that means the AI can perform the action a_t from the current state, s_t to reach the next state s_{t+1}. If a cell (s_t, a_t) contains a 0, that means the AI cannot perform the action a_t from the current state s_t to reach any next state s_{t+1}.

Now, you might remember this very important question, the answer of which is at the heart of Reinforcement Learning.

How will you let the AI know that it has to go to that top priority location **G**?

Everything works with the reward.

I must insist, again, that you remember this. If you attribute a high reward to location **G**, then the AI, through the Q-learning process, will learn to catch that high reward in the most efficient way because it is larger than the rewards of getting to the other locations.

Remember this very important rule: the AI, when it is powered by Q-learning (or deep Q-learning, as you'll soon learn), will always learn to reach the highest reward by the quickest route that does not penalize the AI with negative rewards. That's why the trick to reach location **G** is simply to attribute it a higher reward than the other locations.

Start by manually putting a high reward, which can be any high number as long as it is larger than 1, inside the cell corresponding to location **G**; location **G** is the top priority location where the robot has to go in order to collect the products.

Since location **G** has encoded index state 6, put a 1000 reward in the cell of row 6 and column 6. Later on, we will improve your solution by implementing an automatic way of going to the top priority location, without having to manually update the matrix of rewards and leaving it initialized with 0s and 1s just as it should be. For now, here's your matrix of rewards, including the manual update.

```
# Defining the rewards
R = np.array([[0,1,0,0,0,0,0,0,0,0,0,0],
              [1,0,1,0,0,1,0,0,0,0,0,0],
              [0,1,0,0,0,0,1,0,0,0,0,0],
              [0,0,0,0,0,0,0,1,0,0,0,0],
              [0,0,0,0,0,0,0,0,1,0,0,0],
              [0,1,0,0,0,0,0,0,0,1,0,0],
              [0,0,1,0,0,0,1000,1,0,0,0,0],
              [0,0,0,1,0,0,1,0,0,0,0,1],
              [0,0,0,0,1,0,0,0,0,1,0,0],
              [0,0,0,0,0,1,0,0,1,0,1,0],
              [0,0,0,0,0,0,0,0,0,0,1,0,1],
              [0,0,0,0,0,0,0,1,0,0,1,0]])
```

That completes this first part. Now, let's begin the second part of your implementation.

Part 2 – Building the AI Solution with Q-learning

To build your AI solution, follow the Q-learning algorithm exactly as it was provided previously. If you had any trouble when you tried implementing Q-learning on your own, now is your chance for revenge. Literally, all that's about to follow is only and exactly the same Q-learning process translated into code.

Now you've got that in your mind, try coding it on your own again. You can do it!

Congratulations if you tried, no matter how it came out. Next, let's check if you got it right.

First, initialize all the Q-values by creating your matrix of Q-values full of 0s, in which the rows correspond to the current states s_t, the columns correspond to the actions a_t leading to the next state s_{t+1}, and the cells contain the Q-values $Q(s_t, a_t)$.

```
# PART 2 - BUILDING THE AI SOLUTION WITH Q-LEARNING

# Initializing the Q-values
Q = np.array(np.zeros([12,12]))
```

Then implement the Q-learning process with a for loop over 1,000 iterations, repeating the exact same steps of the Q-learning process 1,000 times.

```
# Implementing the Q-Learning process
for i in range(1000):
    current_state = np.random.randint(0,12)
    playable_actions = []
    for j in range(12):
        if R[current_state, j] > 0:
            playable_actions.append(j)
    next_state = np.random.choice(playable_actions)
    TD = R[current_state, next_state] + gamma * Q[next_state,
np.argmax(Q[next_state,])] - Q[current_state, next_state]
    Q[current_state, next_state] = Q[current_state, next_state] +
alpha * TD
```

Now you've reached the first really exciting step of the journey. You're actually ready to launch the Q-learning process and get your final Q-values. Execute the whole code you've implemented so far, and visualize the Q-values with the following simple print statements:

```
print("Q-values:")
print(Q.astype(int))
```

Here's what I got:

```
Q-values:
[[   0 1661    0    0    0    0    0    0    0    0    0    0]
 [1246    0 2213    0    0 1246    0    0    0    0    0    0]
 [   0 1661    0    0    0    0 2970    0    0    0    0    0]
 [   0    0    0    0    0    0    0 2225    0    0    0    0]
 [   0    0    0    0    0    0    0    0  703    0    0    0]
 [   0 1661    0    0    0    0    0    0    0  931    0    0]
 [   0    0 2213    0    0    0 3968 2225    0    0    0    0]
 [   0    0    0 1661    0    0 2968    0    0    0    0 1670]
 [   0    0    0    0  528    0    0    0    0  936    0    0]
 [   0    0    0    0    0 1246    0    0  703    0 1246    0]
 [   0    0    0    0    0    0    0    0    0  936    0 1661]
 [   0    0    0    0    0    0    0 2225    0    0 1246    0]]
```

If you're working on Spyder in Anaconda, then for more visual clarity you can even check the matrix of Q-values directly in Variable Explorer, by double-clicking on Q. Then, to get the Q-values as integers, you can click on **Format** and enter a float formatting of %.0f. You get the following, which is a bit clearer since you can see the indexes of the rows and columns in your Q matrix:

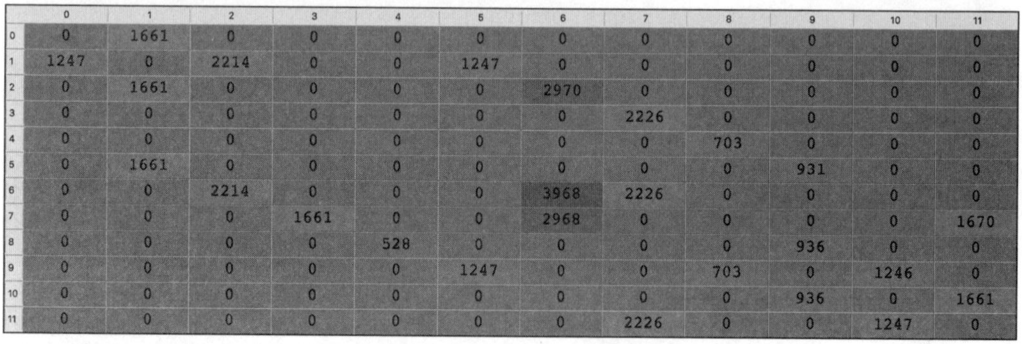

	0	1	2	3	4	5	6	7	8	9	10	11
0	0	1661	0	0	0	0	0	0	0	0	0	0
1	1247	0	2214	0	0	1247	0	0	0	0	0	0
2	0	1661	0	0	0	0	2970	0	0	0	0	0
3	0	0	0	0	0	0	0	2226	0	0	0	0
4	0	0	0	0	0	0	0	0	703	0	0	0
5	0	1661	0	0	0	0	0	0	0	931	0	0
6	0	0	2214	0	0	0	3968	2226	0	0	0	0
7	0	0	0	1661	0	0	2968	0	0	0	0	1670
8	0	0	0	0	528	0	0	0	0	936	0	0
9	0	0	0	0	0	1247	0	0	703	0	1246	0
10	0	0	0	0	0	0	0	0	0	936	0	1661
11	0	0	0	0	0	0	0	2226	0	0	1247	0

Figure 7: Matrix of Q-values

Now that you have your matrix of Q-values, you're ready to go into production—you can move on to the third part of the implementation.

Part 3 – Going into production

In other words, you're going into inference mode! In this part, you'll compute the optimal path from any starting location to any ending top priority location. The idea here is to implement a `route` function, that takes as inputs a starting location and an ending location and that returns as output the shortest route inside a Python list. The starting location corresponds to wherever our autonomous warehouse robot is at a given time, and the ending location corresponds to where the robot has to go as a top priority.

Since you'll want to input the locations with their names (in letters), as opposed to their states (in indexes), you'll need a dictionary that maps the location states (in indexes) to the location names (in letters). That's the first thing to do here in this third part, using a trick to invert your previous dictionary, `location_to_state`, since you simply want to get the exact inverse mapping from this dictionary:

```
# PART 3 - GOING INTO PRODUCTION

# Making a mapping from the states to the locations
state_to_location = {state: location for location, state in location_
to_state.items()}
```

Now, please focus— if the dots haven't perfectly connected in your mind, now is the time when they will. I'll show you the exact steps of how the robot manages to figure out the shortest route.

Your robot is going to go from location **E** to location **G**. Here's the explanation of exactly how it does that—I'll enumerate the different steps of the process. Follow along on the matrix of Q-values as I explain:

1. The AI starts at the starting location **E**.
2. The AI gets the state of location **E**, which according to your `location_to_state` mapping is $s_0 = 4$.
3. On the row of index $s_0 = 4$ in our matrix of Q-values, the AI chooses the column that has the maximum Q-value (703).
4. This column has index 8, so the AI performs the action of index 8, which leads it to the next state $s_{t+1} = 8$.
5. The AI gets the location of state 8, which according to our `state_to_location` mapping is location **I**. Since the next location is location **I, I** is appended to the AI's list containing the optimal path.

6. Then, starting from the new location **I**, the AI repeats the same previous five steps until it reaches our final destination, location **G**.

That's it! That's exactly what you have to implement. You have to generalize this to any starting and ending locations, and the best way to do that is through a function taking two inputs:

1. `starting_location`: The location at which the AI starts
2. `ending_location`: The top priority location to which it has to go

and returning the optimal route. Since we're talking about a route, you can call that function `route()`.

An important thing to understand inside this `route()` function is that since you don't know how many locations the AI will have to go through between the starting and ending locations, you have to make a `while` loop which will repeat the 5-step process described previously, and that will stop as soon as it reaches the top priority end location.

```python
# Making the final function that will return the optimal route
def route(starting_location, ending_location):
    route = [starting_location]
    next_location = starting_location
    while (next_location != ending_location):
        starting_state = location_to_state[starting_location]
        next_state = np.argmax(Q[starting_state,])
        next_location = state_to_location[next_state]
        route.append(next_location)
        starting_location = next_location
    return route
```

Congratulations! Your AI is now ready. Not only does it have the training process implemented, but also the code to run in inference mode. The only thing that's not great so far is that you still have to manually update the matrix of rewards; but no worries, we'll get to that later on. Before we get to that, let's first check that you have an intermediary victory here, and then we can get to work on improvements.

```python
# Printing the final route
print('Route:')
route('E', 'G')
```

The following is the output:

```
Route:
Out[1]:  ['E', 'I', 'J', 'F', 'B', 'C', 'G']
Out[2]:  ['E', 'I', 'J', 'K', 'L', 'H', 'G']
```

That's perfect—I ran the code twice when testing it to go from E to G, which is why you see the two preceding outputs. The two possible optimal paths were returned: one passing by F, and the other one passing by K.

That's a good start. You have a first version of your AI model that functions well. Now let's improve your AI, and take it to the next level.

You can improve the AI in two ways. Firstly, by automating the reward attribution to the top priority location so that you don't have to do it manually. Secondly, by adding a feature that gives the AI the option to go by an intermediate location before going to the top priority location—that intermediate location should be in the top three priority locations.

In our top priority locations ranking, the second top priority location is location K. Therefore, in order to optimize the warehouse flows, your autonomous warehouse robot must go via location K to collect products on its way to the top priority location G. One way to do this is to have the option to go by an intermediate location in the process of your `route()` function. This is exactly what you'll implement as a second improvement.

First, let's implement the first improvement, the one that automates the reward attribution.

Improvement 1 – Automating reward attribution

The way to do this is in three steps.

Step 1: Go back to the original matrix of rewards, as it was before with only 1s and 0s. Part 1 of the code becomes the following, and will be included in the final code:

```
# PART 1 - BUILDING THE ENVIRONMENT

# Defining the states
location_to_state = {'A': 0,
                     'B': 1,
                     'C': 2,
                     'D': 3,
```

```
            'E': 4,
            'F': 5,
            'G': 6,
            'H': 7,
            'I': 8,
            'J': 9,
            'K': 10,
            'L': 11}


# Defining the actions
actions = [0,1,2,3,4,5,6,7,8,9,10,11]


# Defining the rewards
R = np.array([[0,1,0,0,0,0,0,0,0,0,0,0],
              [1,0,1,0,0,1,0,0,0,0,0,0],
              [0,1,0,0,0,0,1,0,0,0,0,0],
              [0,0,0,0,0,0,0,1,0,0,0,0],
              [0,0,0,0,0,0,0,0,1,0,0,0],
              [0,1,0,0,0,0,0,0,0,1,0,0],
              [0,0,1,0,0,0,1,1,0,0,0,0],
              [0,0,0,1,0,0,1,0,0,0,0,1],
              [0,0,0,0,1,0,0,0,0,1,0,0],
              [0,0,0,0,0,1,0,0,1,0,1,0],
              [0,0,0,0,0,0,0,0,0,1,0,1],
              [0,0,0,0,0,0,0,1,0,0,1,0]])
```

Step 2: In part 2 of the code, make a copy (call it R_new) of your rewards matrix, inside which the route() function can automatically update the reward in the cell of the ending location.

Why do you have to make a copy? Because you have to keep the original matrix of rewards initialized with 1s and 0s for future modifications when you want to go to a new priority location. So, how will the route() function automatically update the reward in the cell of the ending location? That's an easy one: since the ending location is one of the inputs of the route() function, then by using your location_to_state dictionary, you can very easily find that cell and update its reward to 1000. Here's how you do that:

```
# Making a function that returns the shortest route from a starting
to ending location
def route(starting_location, ending_location):
    R_new = np.copy(R)
```

```
    ending_state = location_to_state[ending_location]
    R_new[ending_state, ending_state] = 1000
```

Step 3: You must include the whole Q-learning algorithm (including the initialization step) inside the `route()` function, right after we make that update of the reward in your copy (`R_new`) of the rewards matrix. In your previous implementation, the Q-learning process happened on the original version of the rewards matrix. Now that original version needs to stay as it is, that is, initialized to 1s and 0s only. Therefore, you must include the Q-learning process inside the `route()` function, and make it happen on your copy of the rewards matrix `R_new`, instead of the original rewards matrix `R`. Here's how you do that:

```
# Making a function that returns the shortest route from a starting to
ending location
def route(starting_location, ending_location):
    R_new = np.copy(R)
    ending_state = location_to_state[ending_location]
    R_new[ending_state, ending_state] = 1000
    Q = np.array(np.zeros([12,12]))
    for i in range(1000):
        current_state = np.random.randint(0,12)
        playable_actions = []
        for j in range(12):
            if R_new[current_state, j] > 0:
                playable_actions.append(j)
        next_state = np.random.choice(playable_actions)
        TD = R_new[current_state, next_state] + gamma * Q[next_state,
np.argmax(Q[next_state,])] - Q[current_state, next_state]
        Q[current_state, next_state] = Q[current_state, next_state] +
alpha * TD
    route = [starting_location]
    next_location = starting_location
    while (next_location != ending_location):
        starting_state = location_to_state[starting_location]
        next_state = np.argmax(Q[starting_state,])
        next_location = state_to_location[next_state]
        route.append(next_location)
        starting_location = next_location
    return route
```

Perfect; part 2 is now ready! Here's part 2 of the final code in full:

```
# PART 2 - BUILDING THE AI SOLUTION WITH Q-LEARNING

# Making a mapping from the states to the locations
state_to_location = {state: location for location, state in location_
to_state.items()}

# Making a function that returns the shortest route from a starting to
ending location
def route(starting_location, ending_location):
    R_new = np.copy(R)
    ending_state = location_to_state[ending_location]
    R_new[ending_state, ending_state] = 1000
    Q = np.array(np.zeros([12,12]))
    for i in range(1000):
        current_state = np.random.randint(0,12)
        playable_actions = []
        for j in range(12):
            if R_new[current_state, j] > 0:
                playable_actions.append(j)
        next_state = np.random.choice(playable_actions)
        TD = R_new[current_state, next_state] + gamma * Q[next_state,
np.argmax(Q[next_state,])] - Q[current_state, next_state]
        Q[current_state, next_state] = Q[current_state, next_state] +
alpha * TD
    route = [starting_location]
    next_location = starting_location
    while (next_location != ending_location):
        starting_state = location_to_state[starting_location]
        next_state = np.argmax(Q[starting_state,])
        next_location = state_to_location[next_state]
        route.append(next_location)
        starting_location = next_location
    return route
```

If you execute this new code several times with the start and end points of **E** and **G**, you'll get the same two possible optimal paths as before. You can also play around with the route() function and try out different starting and ending points. Try it out!

Improvement 2 – Adding an intermediate goal

Now, let's tackle the second improvement. There are three possible solutions to the problem of adding the option to go by the intermediate location **K**, the second top priority location. When you see them, you'll understand what I meant when I told you that everything in Reinforcement Learning works by the rewards.

Only one of the solutions works from every starting point, but I'd like to give you all three solutions to help reinforce your intuition. To help with that, here's a reminder of our warehouse layout:

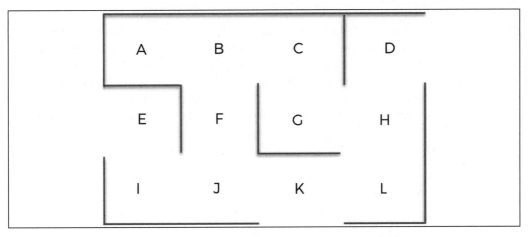

Figure 8: Locations in the warehouse

Solution 1: Give a high reward to the action leading from location **J** to location **K**. This high reward must be larger than 1, and below 1,000. It must be larger than 1 so that the Q-learning process favors the action leading from **J** to **K**, as opposed to the action leading from **J** to **F**, which has a reward of 1. It must also be below 1,000 so that the highest reward stays on the top priority location, to make sure the AI ends up there. For example, in your rewards matrix you can give a high reward of 500 to the cell in the row of index 9 and the column of index 10, since that cell corresponds to the action leading from location **J** (state index 9) to location **K** (state index 10). That way, your AI robot will always go by location **K** when going from location **E** to location **G**. Here's how the matrix of rewards would look in that case:

```
# Defining the rewards
R = np.array([[0,1,0,0,0,0,0,0,0,0,0,0],
              [1,0,1,0,0,1,0,0,0,0,0,0],
              [0,1,0,0,0,0,1,0,0,0,0,0],
              [0,0,0,0,0,0,0,1,0,0,0,0],
              [0,0,0,0,0,0,0,0,1,0,0,0],
              [0,1,0,0,0,0,0,0,0,1,0,0],
              [0,0,1,0,0,0,1,1,0,0,0,0],
              [0,0,0,1,0,0,1,0,0,0,0,1],
              [0,0,0,0,1,0,0,0,0,1,0,0],
              [0,0,0,0,0,1,0,0,1,0,500,0],
              [0,0,0,0,0,0,0,0,0,0,1,0,1],
              [0,0,0,0,0,0,0,1,0,0,1,0]])
```

This solution does not work in every case, and actually only works for starting points **E**, **I**, and **J**. That's because the `500` weight can only affect the decision of the AI as to whether or not it should go from **J** to **K**; it doesn't change how likely it is for the AI to go to **J** in the first place.

Solution 2: Give a bad reward to the action leading from location **J** to location **F**. This bad reward just has to be below 0. By punishing this action with a bad reward, the Q-learning process will never favor the action leading from **J** to **F**. For example, in your rewards matrix, you can give a bad reward of `-500` to the cell in the row of index 9 and the column of index 5, since that cell corresponds to the action leading from location **J** (state index 9) to location **F** (state index 5). That way, your autonomous warehouse robot will never go from location **J** to location **F** on its way to location **G**. Here's how the matrix of rewards would look in that case:

```
# Defining the rewards
R = np.array([[0,1,0,0,0,0,0,0,0,0,0,0],
              [1,0,1,0,0,1,0,0,0,0,0,0],
              [0,1,0,0,0,0,1,0,0,0,0,0],
              [0,0,0,0,0,0,0,1,0,0,0,0],
              [0,0,0,0,0,0,0,0,1,0,0,0],
              [0,1,0,0,0,0,0,0,0,1,0,0],
              [0,0,1,0,0,0,1,1,0,0,0,0],
              [0,0,0,1,0,0,1,0,0,0,0,1],
              [0,0,0,0,1,0,0,0,0,1,0,0],
              [0,0,0,0,0,-500,0,0,1,0,1,0],
              [0,0,0,0,0,0,0,0,0,1,0,1],
              [0,0,0,0,0,0,0,1,0,0,1,0]])
```

This solution does not work in every case, and actually only works for starting points **E**, **I**, and **J**. Just as in solution 1, that's because the `-500` weight can only affect the decision of the AI as to whether or not it should go from **J** to **F**; it doesn't change how likely it is for the AI to go to **J** in the first place.

Solution 3: Make an additional `best_route()` function, taking as inputs the three starting, intermediary, and ending locations, which will call your previous `route()` function twice; the first time from the starting location to the intermediary location, and a second time from the intermediary location to the ending location.

The first two solutions are easy to implement manually, but tricky to implement automatically. It is easy to automatically get the index of the intermediary location via which you want the AI to go, but it's difficult to get the index of the location that leads to that intermediary location, since it depends on the starting location and ending location. If you try to implement either the first or second solution, you'll see what I mean. Besides, solutions 1 and 2 do not work as global solutions.

Only solution 3 guarantees that the AI will visit an intermediate location before going to the final location.

Accordingly, we'll implement solution 3, which can be coded in just two extra lines of code, and which I included in *Part 3 – Going into production*:

```
# PART 3 - GOING INTO PRODUCTION

# Making the final function that returns the optimal route
def best_route(starting_location, intermediary_location, ending_
location):
    return route(starting_location, intermediary_location) +
route(intermediary_location, ending_location)[1:]

# Printing the final route
print('Route:')
best_route('E', 'K', 'G')
```

Easy, right? Sometimes, the best solutions are the simplest ones. That's definitely the case here. As you can see, included in Part 3 is the code that runs the ultimate test. This test will be successful if the AI goes through location **K** while taking the shortest route from location **E** to location **G**. To test it, execute this whole new code as many times as you want; you'll always get the same, expected output:

```
Route:
['E', 'I', 'J', 'K', 'L', 'H', 'G']
```

Congratulations! You've developed a fully functional AI, powered by Q-learning, which solves an optimization problem for logistics. Using this AI robot, we can now go from any location to any new top priority location, while optimizing our paths to collect products in a second priority intermediary location. Not bad! If you get bored with logistics, feel free to imagine yourself back in the maze, and try the `best_route()` function with whatever starting and ending points you would like, so you can see how flexible the AI you've created is. Have fun with it! And, of course, you have the full code available for you on the GitHub page.

Summary

In this chapter, you've implemented a Q-learning solution to a business problem. You had to find the best route to a certain location in your warehouse. Not only have you done that, but you've also implemented additional code that allowed your AI to make as many intermediary stops as you wanted. Based on the obtained rewards, your AI was able to find the best route going through these stops. That was Q-learning for warehouse robots. Now, let's move on to deep Q-learning!

9
Going Pro with Artificial Brains – Deep Q-Learning

This next AI model is fantastic, because it is the first AI model that is really inspired by human intelligence. I hope you're ready to go pro on the next exciting step in your AI journey; this book is not only a crash course on AI, but also an introduction to deep learning.

Today, some of the top AI models integrate deep learning. They form a new branch of AI called deep Reinforcement Learning. The model we'll cover in this chapter belongs to that branch, and is called deep Q-learning. You already know what Q-learning is all about, but you might not know anything about deep learning and **Artificial Neural Networks** (**ANNs**); we'll start with them. Of course, if you are an expert in deep learning, you can skip the first sections of this chapter, but consider that a little refresher never hurt anyone.

Before we start going through the theory, you'll begin with real, working code written in Python. You'll create some AI first, and then I'll help you understand it afterwards. Right now, we're going to build an ANN to predict house prices.

Predicting house prices

What we want to do is predict how much a certain house might cost, based on some variables. In order to do so you need to follow these four steps:

1. Get some historical data on house sales; for this example, you'll use a dataset of about 20,000 houses in Seattle.

2. Import this data to your code while applying some scaling to your variables (I'll explain scaling to you as we go).

3. Build an Artificial Neural Network using any library—you'll use Keras, as it is simple and reliable.

4. Train your ANN and get the results.

Now that you know the structure of your future code, you can start writing it. Since all the libraries that you'll use are available in Google Colab, you can easily use it to perform this task.

Uploading the dataset

Start by creating a new Google Colab notebook. Once we have created your new notebook, before you start coding anything, you have to upload your dataset. You can find this dataset, called `kc_house_data.csv`, on the GitHub repository in the `Chapter 09` folder.

Figure 1: GitHub – Chapter 09

Once you have done that, you can upload it to Colab by doing the following:

1. Click this little arrow here:

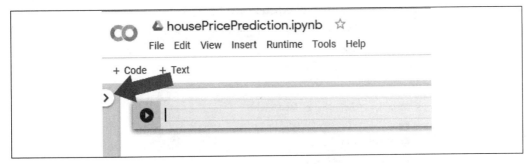

Figure 2: Google Colab – Uploading files (1/3)

2. In the window that pops up, go to **Files**. You should get something like this:

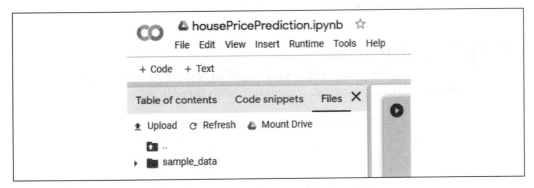

Figure 3: Google Colab – Uploading files (2/3)

3. Click on **UPLOAD** and then select the file location where you saved the `kc_house_data` dataset.

4. After you have done that, you should get a new folder with our dataset, like this:

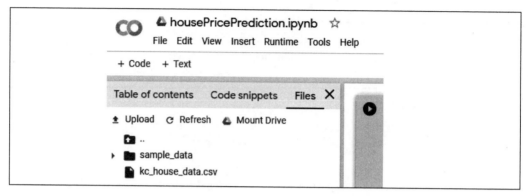

Figure 4: Google Colab – Uploading files (3/3)

Great! Now you can start coding.

Importing libraries

Every time you start coding something you ought to begin by importing the necessary libraries. Therefore, we start our code with these lines:

```
3    # Importing the libraries
4    import pandas as pd
5    import numpy as np
6    import keras
7    from sklearn.model_selection import train_test_split
8    from sklearn.preprocessing import MinMaxScaler
9    from keras.layers import Dense, Dropout
10   from keras.models import Sequential
11   from keras.optimizers import Adam
```

In lines 4 and 5, after the comment, you import the pandas and numpy libraries. Pandas will help you read the dataset and NumPy is very useful when you're dealing with arrays or lists; you'll use it to drop some unnecessary columns from your dataset.

In the two subsequent lines you import two useful tools from the Scikit-Learn library. The first one is a tool that will help split the dataset into a training set and a test set (you should always have both of them; the AI model is trained on the training set and then tested on the test set) and the second one is a scaler that will help you later when scaling values.

Lines 9, 10, and 11 are responsible for importing the keras library, which you'll use in order to build a neural network. Each of these tools is used later in the code.

Now that you have imported your libraries you can read the dataset. Do it by using the Pandas library you imported before, with this one line:

```
13      # Importing the dataset
14      dataset = pd.read_csv('kc_house_data.csv')
```

Since you used pd as an abbreviation for the Pandas library when you imported it, you can use it to shorten your code. After you call the Pandas library with pd, you can use one of its functions, read_csv, which, as the name suggests, reads csv files. Then in the brackets you input the file name, which in your case is kc_house_data. csv. No other arguments are needed.

Now I have a little exercise for you! Have a look at the dataset and try to judge which of the variables will matter for our price prediction. Believe me, not all of them are relevant. I strongly suggest that you try to do it alone even though we'll discuss them in the next section.

Excluding variables

Were you able to discern which variables are necessary and which are not? Don't worry if not; we'll explain them and their relevance right now.

The following table explains every column in our dataset:

Variable	Description
Id	Unique ID for each household
Date	Date when the house was sold
Price	How much the house cost when sold
Bedrooms	Number of bedrooms
Bathrooms	Number of bathrooms; 0.5 represents room with a toilet but no shower
Sqft_living	Square footage of the apartment's interior living space
Sqft_lot	Square footage of the land space
Floors	Number of floors
Waterfront	0 if the apartment doesn't overlooking the waterfront, 1 if it does

Variable	Description
View	Value in the range 0-4 depending on how good the view of the property is
Condition	Value from 1-5 defining the condition of the property
Grade	Value from 1-13 indicating the design and construction of the building
Sqft_above	The square footage of the interior housing space that is above ground level
Sqft_basement	The square footage of the basement
Yr_built	Year when the house was built
Yr_renovated	Year when the house was renovated (0 if wasn't)
Zipcode	Zip code of the area house is located in
Lat	Latitude
Long	Longitude
Sqft_living15	The square footage of the interior housing living space for the nearest 15 neighbors
Sqft_lot15	Square footage of the land lots of the nearest 15 neighbors

It turns out that from those 21 variables, only 18 count. That is because unique, category-like values do not have any impact on your prediction. That includes Id, Date, and Zipcode. Price is the target of your prediction, and therefore you should get rid of that from your variables as well. After all that, you have 17 independent variables.

Now that we have explained all the variables and decided which are relevant and which are not, you can go back to your code. You're going to exclude these unnecessary variables and split the dataset into the features and the target (in our case the target is price).

```
16    # Getting separately the features and the targets
17    X = dataset.iloc[:, 3:].values
18    X = X[:, np.r_[0:13,14:18]]
19    y = dataset.iloc[:, 2].values
```

On line 17, you take all rows and all columns starting with the fourth one (since you're excluding Id, Date, Price) from your dataset and call this new set x. You use .iloc to slice the dataset, and then take .values to change it to a NumPy object. These will be your features.

Next you need to exclude Zipcode, which quite unfortunately is in the middle of the features set. That's why you have to use a NumPy function (np.r_) that separates x, excludes the columns you choose (in this case it is column 14. 13 is the index of this column, since indexes in Python start with zero; it's also worth mentioning that upper bounds are excluded in Python notation, which is why we write 0:13), and then connects them once again to form a new array. In the next line, you get the target of your prediction and call it y. This corresponds to the third column in your dataset, that is, Price.

Data preparation

Now that you've separated your important features and target, you can split your x and y into training and test sets. We do that with the following line:

```
21    # Splitting the dataset into a training set and a test set
22    X_train, X_test, y_train, y_test = train_test_split(X, y, test_
      size = 0.2, random_state = 0)
```

This is very important when doing any kind of machine learning. You always have to have a training set on which you train your model, and a test set on which you test it. You perform that operation using the train_test_split function you imported before. After doing that, you get X_train, which is of equal size to y_train, and each of them are exactly 80% of our previous X and y set. X_test and y_test are made up of the remaining 20% of X and y.

Now that you have both a training set and a test set, what do you think the next step is? Well, you have to scale your data.

Scaling data

Now you might be wondering why on earth you have to perform such an operation. You already have the data, so why not build and train the neural network already?

There's a problem with that; if we leave the data as it is, you'll notice that your ANN does not learn. The reason for that is because different variables will impact your prediction more or less depending on their values.

Take this graph illustrating what I mean, based on a property that has 3 bedrooms and 1,350 square feet of living area.

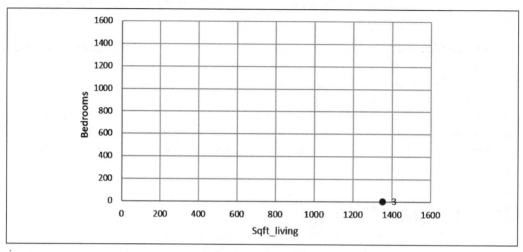

Figure 5: Example for 3 bedrooms and 1350 square feet of living area

You can clearly see that the number of bedrooms won't affect the prediction as much as Sqft_living will. Even we humans cannot see any difference between zero bedrooms and three bedrooms on this graph.

One of many solutions to this problem is to scale all variables to be in a range between 0 and 1. We achieve this by calculating this equation:

$$x_{scaled} = \frac{x - x_{min}}{x_{max} - x_{min}}$$

where:

- x – the value we are scaling in our case every value in a column
- x_{min} – minimum value across all in a column
- x_{max} – maximum value across all in a column
- x_{scaled} – x after performing scaling

After performing this scaling, our previous graph now looks something like this:

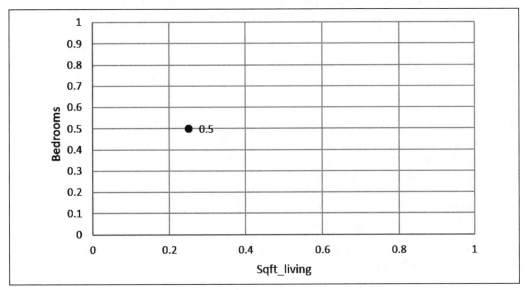

Figure 6: Same graph after scaling

Now we can undoubtedly say that the number of bedrooms will have a similar impact to Sqft_living. We can clearly see the difference between zero bedrooms and three bedrooms.

So, how do we implement that in code? Since you know the equation, I recommend that you try to do it yourself. Don't worry if you fail; I'll show you a very simple way to do it in the next paragraph.

If you were able to scale the data on your own, then congratulations! If not, follow along through this next section to see the answer. You might have noticed that you imported a class of Scikit-learn library called `MinMaxScaler`. You can use that class to scale the variables with the following code:

```
24      # Scaling the features
25      xscaler = MinMaxScaler(feature_range = (0,1))
26      X_train = xscaler.fit_transform(X_train)
27      X_test = xscaler.transform(X_test)
28
29      # Scaling the target
30      yscaler = MinMaxScaler(feature_range = (0,1))
31      y_train = yscaler.fit_transform(y_train.reshape(-1,1))
32      y_test = yscaler.transform(y_test.reshape(-1,1))
```

This code creates two scalers, one to scale the features and one to scale the targets. Call them xscaler and yscaler. The feature_range argument is the range to which you want your data to be scaled (from 0 to 1 in your case).

Then you use the fit_transform method, which scales X_train and y_train and adjusts the scalers based on these sets (fit part of this method sets x_{min} and x_{max}). After that you use the transform method to scale X_test and y_test without adjusting yscaler and xscaler.

When scaling the y variables, you have to reshape them by using .reshape(-1,1) in order to create a fake second dimension (so the code can treat this one-dimensional array as a two-dimensional array with one column). We need this fake second dimension to avoid a format error.

If you still do not understand why we have to use scaling, please read this section once again. It'll also get clearer once we go through the theory.

Finally, you can proceed to building a neural network! Keep in mind that all the theory behind it will be covered later in the chapter, so don't be scared if you have trouble understanding something.

Building the neural network

To build the neural network, you can use a highly reliable and easy to use library called Keras. Let's get straight into coding it:

```
34       # Building the Artificial Neural Network
35       model = Sequential()
36       model.add(Dense(units = 64, kernel_initializer = 'uniform',
         activation = 'relu', input_dim = 17))
37       model.add(Dense(units = 16, kernel_initializer = 'uniform',
         activation = 'relu'))
38       model.add(Dense(units = 1, kernel_initializer = 'uniform',
         activation = 'relu'))
39       model.compile(optimizer = Adam(lr = 0.001), loss = 'mse',
         metrics = ['mean_absolute_error'])
```

In line 35 of the code block you instantiate your model by using the Sequential class from the Keras library.

Next, you add a line that adds a new layer with 64 neurons to your neural network. kernel_initializer is an argument that defines the way the initial weights are created in the layer, activation is the activation function of this layer and input_dim is the size of the input; in your case, these are the 17 features that define how much a house costs.

Next, you add two more layers, one with 16 neurons and one with 1 neuron that will be the output of the neural network.

In the final line of this snippet you use the `compile` method, which describes how you want to train your net. Inside this `compile` method, `optimizer` is the tool that performs backpropagation, `lr` is the learning rate — the speed at which the weights in the ANN are updated. `loss` is how you want to calculate the error of the output (I have decided to go for the mean squared error `mse`), and `metrics` is just a value that will help you visualize performance — you can use mean absolute error.

If you don't know what I'm talking about right now, what activations, losses, and optimizers are, you don't have to worry. You'll understand them soon, when we get to the theory later in the chapter.

Training the neural network

Now that you've built your model, you can finally train it!

```
41      # Training the Artificial Neural Network
42      model.fit(X_train, y_train, batch_size = 32, epochs = 100,
        validation_data = (X_test, y_test))
```

This simple one-liner is responsible for learning.

As the first two arguments of this fit method, you input `X_train` and `y_train` which are the sets your model will be trained on. Then you have an argument called `batch_size`; this defines after how many records in your dataset you update your weights (loss is summed up and back-propagated after `batch_size` inputs). Next you have `epochs`, and this value defines how many times you teach your model on the entire `X_train` and `y_train` set. The final argument is `validation_data`, and there, as you can see, you put `X_test` and `y_test`. This means that after every epoch, your model will be tested on this set, but it won't learn from it.

Displaying results

You're nearly there; you have just one last non-obligatory step to take. You calculate the absolute error on the test set and see its real, unscaled predictions (actual prices, not in the range (0,1)).

```
44      # Making predictions on the test set while reversing the scaling
45      y_test = yscaler.inverse_transform(y_test)
46      prediction = yscaler.inverse_transform(model.predict(X_test))
47
```

```
48        # Computing the error rate
49        error = abs(prediction - y_test)/y_test
50        print(np.mean(error))
```

You rescale back your `y_test` on line 45. Then, you make a prediction on your test set of features and rescale it back too, since the predictions are also scaled down.

In the last two lines you calculate the absolute error using the formula:

$$Error = \frac{\left|prediction - actualValue\right|}{actualValue} * 100\%$$

Since both prediction and `y_test` are NumPy arrays, you can divide them by simply using the / symbol. In the last line, you calculate the mean error using a NumPy function.

Superb! Now that you have it all finished, you can finally run this code and see the results.

```
17290/17290 [==============] - 1s 42us/step - loss: 2.3576e-04 - mean_absolute_error: 0.0096 - val_loss: 2.4573e-04
Epoch 90/100
17290/17290 [==============] - 1s 42us/step - loss: 2.3229e-04 - mean_absolute_error: 0.0096 - val_loss: 2.5893e-04
Epoch 91/100
17290/17290 [==============] - 1s 42us/step - loss: 2.2763e-04 - mean_absolute_error: 0.0095 - val_loss: 2.9130e-04
Epoch 92/100
17290/17290 [==============] - 1s 42us/step - loss: 2.2835e-04 - mean_absolute_error: 0.0096 - val_loss: 2.8402e-04
Epoch 93/100
17290/17290 [==============] - 1s 43us/step - loss: 2.3680e-04 - mean_absolute_error: 0.0097 - val_loss: 2.5020e-04
Epoch 94/100
17290/17290 [==============] - 1s 43us/step - loss: 2.3185e-04 - mean_absolute_error: 0.0097 - val_loss: 2.5359e-04
Epoch 95/100
17290/17290 [==============] - 1s 42us/step - loss: 2.3206e-04 - mean_absolute_error: 0.0096 - val_loss: 2.5415e-04
Epoch 96/100
17290/17290 [==============] - 1s 40us/step - loss: 2.3863e-04 - mean_absolute_error: 0.0097 - val_loss: 2.7108e-04
Epoch 97/100
17290/17290 [==============] - 1s 38us/step - loss: 2.2833e-04 - mean_absolute_error: 0.0097 - val_loss: 2.5300e-04
Epoch 98/100
17290/17290 [==============] - 1s 40us/step - loss: 2.2590e-04 - mean_absolute_error: 0.0096 - val_loss: 2.4964e-04
Epoch 99/100
17290/17290 [==============] - 1s 40us/step - loss: 2.2953e-04 - mean_absolute_error: 0.0096 - val_loss: 2.8073e-04
Epoch 100/100
17290/17290 [==============] - 1s 38us/step - loss: 2.2992e-04 - mean_absolute_error: 0.0095 - val_loss: 2.5800e-04
13.446597905432984
```

Figure 7: Results

As you can see in the last line, your result is shown. In my case the average error was 13.5%. That is a really good result!

Now we can get into the theory behind deep learning, and find out how a neural network really works.

Deep learning theory

Here is our plan of attack to go pro and tackle deep learning:

1. The neuron
2. The activation function
3. How do neural networks work?
4. How do neural networks learn?
5. Forward-propagation and back-propagation
6. Gradient descent, including Batch, Stochastic, and Mini-Batch methods

I hope you're excited about this section—deep learning is an awesome and powerful field to study.

The neuron

The neuron is the basic building block of Artificial Neural Networks, and they are based on the neuron cells found the brain.

Biological neurons

In the following images are real-life neurons that have been smeared onto a slide, colored a little bit, and observed through a microscope:

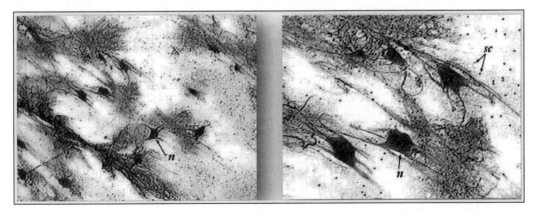

Figure 8: The neuron

As you can see, they have the structure of a central body with lots of different branches coming out of it. The question is: How can we recreate that in a machine? We really want to recreate it in a machine, since the whole purpose of deep learning is to mimic how the human brain works in the hope that by doing so we create something amazing: a powerful infrastructure for learning machines.

Why do we hope for that? Because the human brain just happens to be one of the most powerful learning tools on the planet. We hope that if we recreate it, then we'll have something just as awesome as that.

Our challenge right now, our very first step in creating artificial neural networks, is to recreate a neuron. So how do we do it? Well, first of all let's take a closer look at what a neuron actually is.

In 1899, the neuroscientist Santiago Ramón y Cajal dyed neurons in actual brain tissue, and looked at them under a microscope. While he was looking at them, he drew what he saw, which was something very much like the slides we looked at before. Today, technology has advanced quite a lot, allowing us to see neurons much more closely and in more detail. That means that we can draw what they look like diagrammatically:

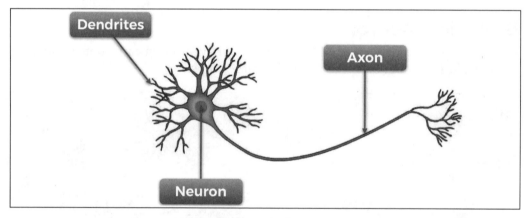

Figure 9: The neuron's structure

This neuron exchanges signals between its neighbor neurons. The dendrites are the receivers of the signal and the axon is the transmitter of the signal.

The dendrites of the neuron are connected to the axons of other neurons above it. When the neuron fires, the signal travels down its axon and passes on to the dendrites of the next neuron. That is how they are connected, and how a neuron works. Now we can move from neuroscience to technology.

Artificial neurons

Here's how a neuron is represented inside an Artificial Neural Network:

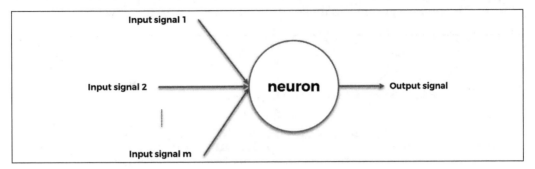

Figure 10: An Artificial Neural Network with a single neuron

Just like a human neuron, it gets some input signals and it has an output signal. The blue arrow connecting the input signals to the neuron, and the neuron to the output signal, are like the synapses in the human neuron.

Here in the artificial neuron, what exactly are the input and output signals going to be? The input signals are the scaled independent variables composing the states of the environment. For example, in the server cooling practical example we'll code later in this book (*Chapter 11, AI for Business – Minimize Costs with Deep Q-Learning*), these are the temperature of the server, the number of users, and the rate of data transmission. The output signal is the output values, which in a deep Q-learning model are always the Q-Values. Knowing all that, we can make a general representation of a neuron for machines:

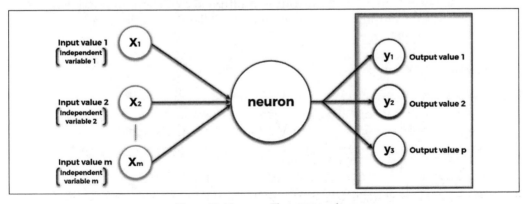

Figure 11: Neuron – The output values

To finish describing the neuron, we need to add the last element missing from this representation, which is also the most important one: the weights.

Each synapse (blue arrow) is attributed a weight. The larger the weight, the stronger the signal is through the synapse. What is fundamental to understand is that these weights are what the machine updates over time to improve its predictions. Let's add them to the previous graphic, to make sure you can visualize them well:

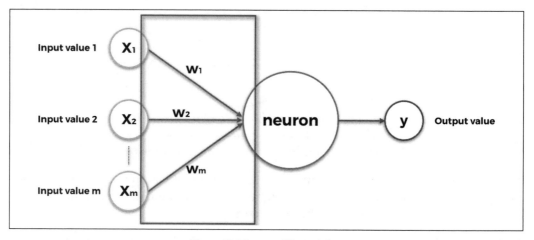

Figure 12: Neuron – The weights

That's the neuron. The next thing to understand is the activation function; the way the neuron decides what output to produce given a set of inputs.

The activation function

The activation function is the function ϕ, operating inside the neuron, that takes as inputs the linear sum of the input values multiplied by their associated weights, and that returns the output value as shown in the following graphic:

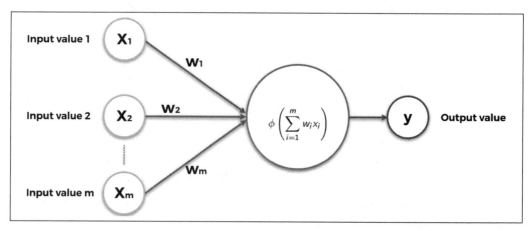

Figure 13: The activation function

such that:

$$y = \phi\left(\sum_{i=1}^{m} w_i x_i\right)$$

Your next question is probably: what exactly is the function ϕ?

There can be many of them, but here we'll describe the three most used ones, including the one you'll use in the practical activity:

1. The threshold activation function
2. The sigmoid activation function
3. The rectifier activation function

Let's push your expertise further by having a look at them one by one.

The threshold activation function

The threshold activation function is simply defined by the following:

$$\phi(x) = 1 \text{ if } x \geq 0$$

$$\phi(x) = 0 \text{ if } x < 0$$

and can be represented by the following curve:

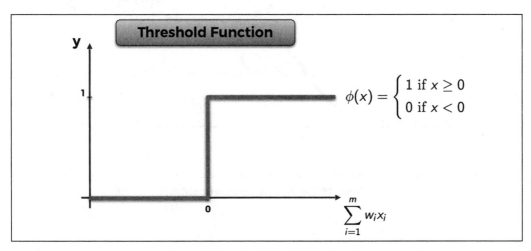

Figure 14: The threshold activation function

This means that the signal passing through the neuron is discontinuous, and will only be activated if:

$$\sum_{i=1}^{m} w_i x_i \geq 0$$

Now let's have a look at the next activation function: the sigmoid activation function. The sigmoid activation function is the most effective and widely used one in Artificial Neural Networks, but mostly in the last hidden layer that leads to the output layer.

The sigmoid activation function

The sigmoid activation function is defined by the following:

$$\phi(x) = \frac{1}{1+e^{-x}}$$

and can be represented by the following curve:

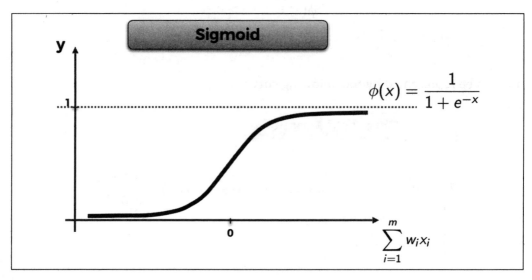

Figure 15: The sigmoid activation function

This means that the signal passing through the neuron is continuous and will always be activated. And the higher the value of:

$$\sum_{i=1}^{m} w_i x_i$$

the stronger the signal.

Now let's have a look at another widely used activation function: the rectifier activation function. You'll find it in most of the deep neural networks, but mostly inside the early hidden layers, as opposed to the sigmoid function, which is rather used for the last hidden layer leading to the output layer.

The rectifier activation function

The rectifier activation function is simply defined by the following:

$$\phi(x) = \max(x, 0)$$

and is therefore represented by the following curve:

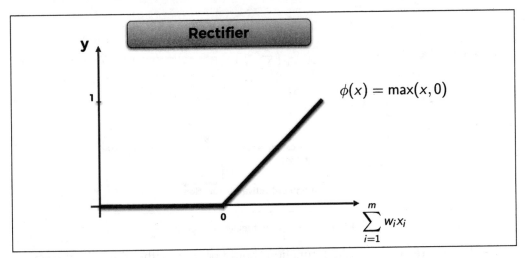

Figure 16: The rectifier activation function

This means that the signal passing through the neuron is continuous, and will only be activated if:

$$\sum_{i=1}^{m} w_i x_i \geq 0$$

The higher the weighted sum of inputs, the stronger the signal.

That raises the question: which activation function should you choose, or, as it's more frequently asked, how do you know which one to choose?

The good news is that the answer is simple. It actually depends on what gets returned as the dependent variable. If it's a binary outcome, 0 or 1, then a good choice would be the threshold activation function. If what you want returned is the probability that the dependent variable is 1, then the sigmoid activation function is an excellent choice, since its sigmoid curve is a perfect fit to model probabilities.

To recap, here's the small blueprint highlighted in this figure:

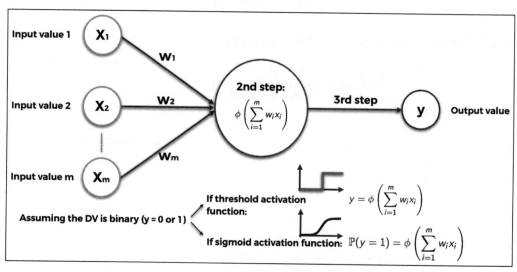

Figure 17: Activation function blueprint

Remember, the rectifier activation function should be used within the hidden layers of a deep neural network with more than one hidden layer, and the sigmoid activation function should be used in the last hidden layer leading to the output layer.

Let's highlight this in the following figure so that you can visualize it and remember it better:

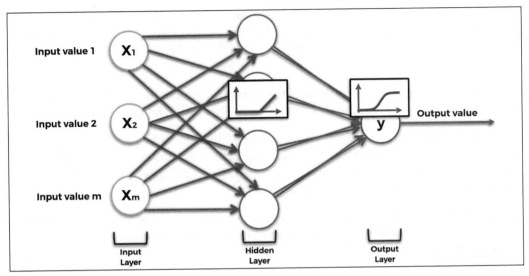

Figure 18: Different activation functions in different layers

We're progressing fast! You already know quite a lot about deep learning. It's not over yet though — let's move on to the next section to explain how neural networks actually work.

How do neural networks work?

To explain this, let's go back to the problem of predicting real estate prices. We had some independent variables which we were using to predict the price of houses and apartments. For simplicity's sake, and to be able to represent everything in a graph, let's say that our only independent variables (our predictors) are the following:

1. Area (square feet)
2. Number of bedrooms
3. Distance to city (miles)
4. Age

Our dependent variable is the apartment price that we're predicting. Here's how the magic works in deep learning.

A weight is attributed to each of the independent, scaled variables in such a way that the higher the weight is, the more of an effect the independent variable will have on the dependent variable; that is, the stronger a predictor it will be of the dependent variable.

As soon as new inputs enter the neural network, the signals are forward-propagated from each of the inputs, reaching the neurons of the hidden layer.

Inside each neuron of the hidden layer, the activation function is applied, so that the lower the weight of the input, the more the activation function blocks the signal coming from that input, and the higher the weight of that input, the more the activation function lets that signal go through.

Finally, all the signals coming from the hidden neurons, more or less blocked by the activation functions, are forward propagated to the output layer, to return the final outcome: the price prediction.

Here's a visualization of how that neural network works:

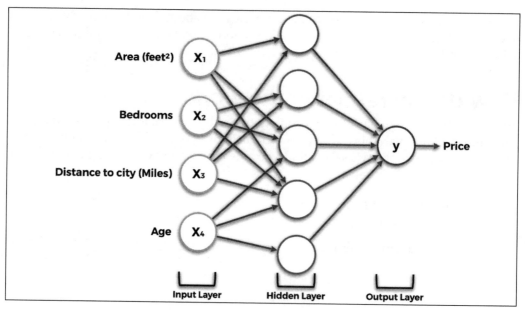

Figure 19: How Neural Networks work – Example in real estate price prediction

That covers half of the story. Now we know how a neural network works, we need to find out how it learns.

How do neural networks learn?

Neural networks learn by updating, over many iterations, the weights of all the inputs and hidden neurons (when having several hidden layers), always towards the same goal: to reduce the loss error between the predictions and the actual values.

In order for neural networks to learn, we need the actual values, which are also called the targets. In our preceding example about real estate pricing, the actual values are the real prices of the houses and apartments taken from our dataset. These real prices depend on the independent variables listed previously (area, number of bedrooms, distance to city, and age), and the neural network learns to make better predictions of these prices, by running the following process:

1. The neural network forward propagates the signals coming from the inputs; independent variables x_1, x_2, x_3 and x_4.

2. Then it gets the predicted price $\hat{y}$ in the output layer.

3. Then it computes the loss error, C, between the predicted price $\hat{y}$ (prediction) and the actual price y (target):

$$C = \frac{1}{2}(\hat{y} - y)^2$$

4. Then this loss error is back-propagated inside the neural network, from right to left in our representation.

5. Then, on each of the neurons, the neural network runs a technique called gradient descent (which we will discuss in the next section) to update the weights in the direction of loss reduction, that is, into new weights which reduce the loss error C.

6. Then this whole process is repeated many times, with each time new inputs and new targets, until we get the desired performance (early stopping) or the last iteration (the number of iterations chosen in the implementation).

Let's show the two main phases, forward-propagation and back-propagation, of this whole process in two separate graphics in the next section.

Forward-propagation and back-propagation

Phase 1: Forward-propagation:

Here's how the signal is forward-propagated throughout the artificial neural network, from the inputs to the output:

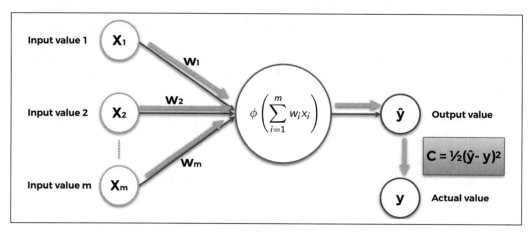

Figure 20: Forward-propagation

Once the signal's been propagated through the entire network, the loss error C is calculated so that it can be back-propagated.

Phase 2: Back-propagation:

And after forward-propagation comes back-propagation, during which the loss error C is propagated back into the neural network from the output to the inputs.

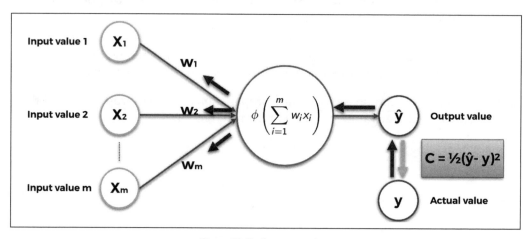

Figure 21: Back-propagation

During back-propagation, the weights are updated to reduce the loss error C between the predictions (output value) and the targets (actual value). How are they updated? This is where gradient descent comes into play.

Gradient Descent

Gradient descent is an optimization technique that helps us find the minimum of a cost function, like the preceding loss error C we had:

$$C = \frac{1}{2}(\hat{y} - y)^2$$

Let's visualize it in the most intuitive way, like the following ball in a bowl (with a little math sprinkled on top):

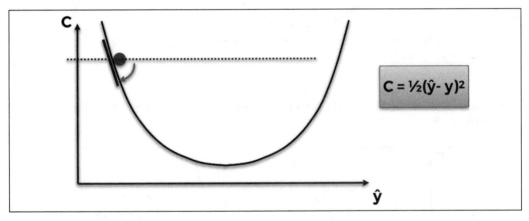

Figure 22: Gradient Descent (1/4)

Imagine this is a cross section of a bowl, into which we drop a small red ball and let it find its way down to the bottom of the bowl. After some time, it will stop rolling, when it finds the sweet spot at the bottom of the bowl.

You can think about gradient descent in the same way. It starts somewhere in the bowl (initial values of parameters) and tries to find the bottom of the bowl, or in other words, the minimum of a cost function.

Let's go through the example that is shown in the preceding image. The initial values of the parameters have set our ball at the position shown. Based on that we get some predictions, which we compare to our target values. The difference between these two sets is our loss for the current set of parameters.

Then we calculate the first derivative of the cost function, with respect to the parameters. This is where the name **gradient** comes from. Here, this first derivative gives us the slope of the tangent to the curve where the ball is. If the gradient of the slope is negative, like on the preceding image, we take the next step to the right side. If the gradient of the slope is positive, we take the next step to the left side.

The name **descent** thus comes from the fact that we always take the next step that points downhill, as represented in the following graphic:

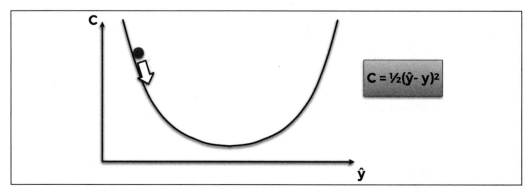

Figure 23: Gradient Descent (2/4)

In the next position our ball rests on a positive slope, so we have to take the next step to the left:

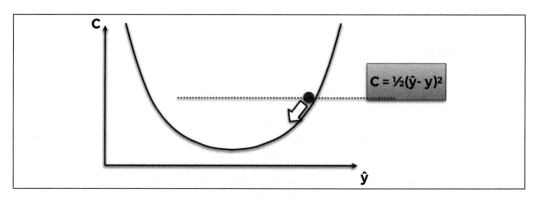

Figure 24: Gradient Descent (3/4)

Eventually, by repeating the same steps, the ball will end up at the bottom of the bowl:

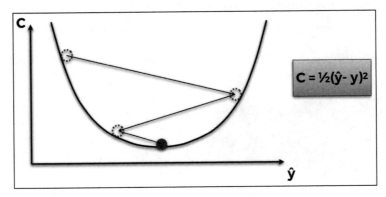

$$C = \frac{1}{2}(\hat{y} - y)^2$$

Figure 25: Gradient Descent (4/4)

And that's it! That's how gradient descent operates in one dimension (one parameter). Now you might ask: "Great, but how does this scale?" We saw an example of one-dimensional optimization, but what about two or even three dimensions?

It's an excellent question. gradient descent guarantees that this approach scales on as many dimensions as needed, provided the cost function is convex. In fact, if the cost function is convex, gradient descent will find the absolute minimum of the cost function. Following is an example in two dimensions:

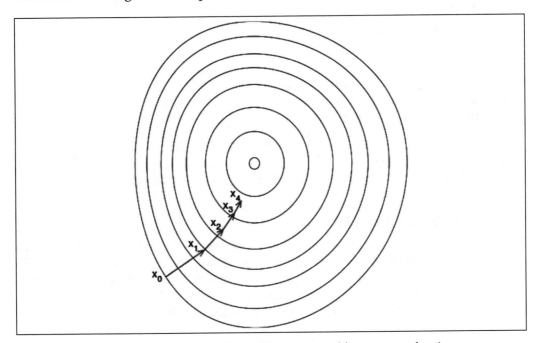

Figure 26: Gradient Descent – Convergence guaranteed for convex cost functions

However, if the cost function is not convex, gradient descent will only find a local minimum. Here is an example in three dimensions:

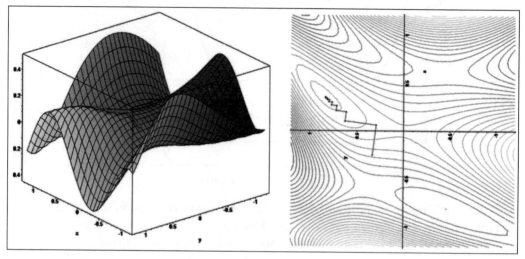

Figure 27: Example of non-convergence (right) for a non-convex function (left)

Now that we understand what gradient descent is all about, we can study the most advanced and most effective versions of it:

1. Batch gradient descent
2. Stochastic gradient descent
3. Mini-batch gradient descent

"Gradient descent", "batch gradient descent", "mini batch gradient descent", "stochastic gradient descent," there are so many terms and someone like you who's just starting may find themselves very confused. Don't worry—I've got your back.

The main difference between all of these versions of gradient descent is just the way we feed our data to a model, and how often we update our parameters (weights) to move our small red ball. Let's start by explaining batch gradient descent.

Batch gradient descent

Batch gradient descent is when we have a batch of inputs (as opposed to a single input) feeding the neural network, forward-propagating them to obtain in the end a batch of predictions, which themselves are compared to a batch of targets. The global loss error between the predictions and the targets of the two batches is then computed as the sum of the loss errors between each prediction and its associated target.

That global loss is back-propagated into the neural network, where gradient descent or stochastic gradient descent is performed to update all the weights, according to how much they were responsible for that global loss error.

Here is an example of batch gradient descent. The problem to solve is about predicting the score (from 0 to 100 %) students get in an exam, based on the time spent studying (Study Hrs) and the time spent sleeping (Sleep Hrs):

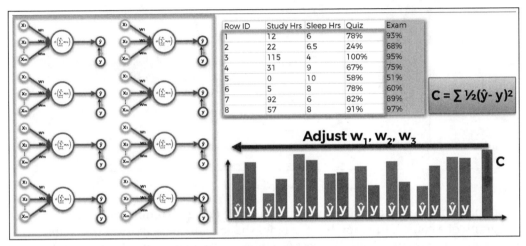

Row ID	Study Hrs	Sleep Hrs	Quiz	Exam
1	12	6	78%	93%
2	22	6.5	24%	68%
3	115	4	100%	95%
4	31	9	67%	75%
5	0	10	58%	51%
6	5	8	78%	60%
7	92	6	82%	89%
8	57	8	91%	97%

$$C = \sum \tfrac{1}{2}(\hat{y} - y)^2$$

Figure 28: Batch Gradient Descent

An important thing to note on this preceding graphic is that these are not multiple neural networks, but a single one represented by separate weight updates. As we can see in this example of batch gradient descent, we feed all of our data into the model at once.

This produces collective updates of the weights and fast optimization of the network. However, there is a bad side to this as well. There is, once again, the possibility of getting stuck in a local minimum, as we can see in the following graphic:

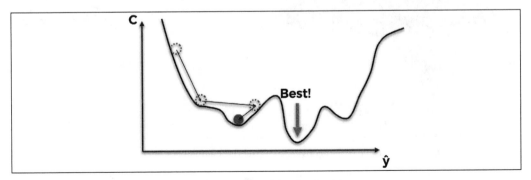

Figure 29: Getting stuck in a local minimum

We explained the reason why this happens a bit earlier: it is because the cost function in the preceding graphic is not convex, and this type of optimization (simple gradient descent) requires the cost function to be convex. If that is not the case, we can find ourselves stuck in a local minimum and never find the global minimum with the optimal parameters. On the other hand, here is an example of a convex cost function, the same one as we saw earlier:

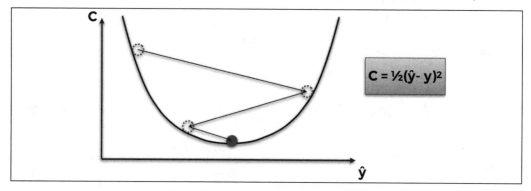

Figure 30: An example of a convex function

In simple terms, a function is convex if it has only one global minimum. And the graph of a convex function has the bowl shape. However, in most problems, including business problems, the cost function will not be convex (as in the following graphic example in 3D), and thus not allow simple gradient descent to perform well. This is where stochastic gradient descent comes into play.

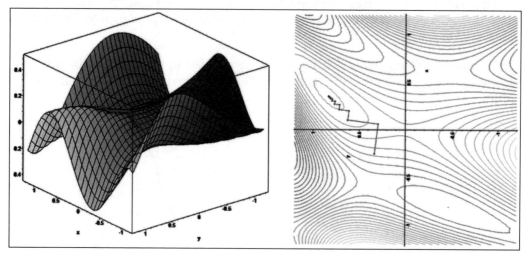

Figure 31: Example of non-convergence (right) for a non-convex function (left)

Stochastic gradient descent

Stochastic Gradient Descent (SGD) comes to save the day. It provides better results overall, preventing the algorithm from getting stuck in a local minimum. However, as its name suggests, it is stochastic, or in other words, random.

Because of this property, no matter how many times you run the algorithm, the process will always be slightly different, regardless of the initialization.

SGD does not run on the whole dataset at once, but instead input by input. The process goes like this:

1. Input a single observation.
2. Forward propagate that input to get a single prediction.
3. Compute the loss error between the prediction (output) and the target (actual value).
4. Back-propagate the loss error into the neural network.
5. Update the weights with gradient descent.
6. Repeat steps 1 to 5 through the whole dataset.

Let's show the first three iterations on the first three single inputs for the example we looked at earlier, predicting the scores in an exam:

First input row of observation:

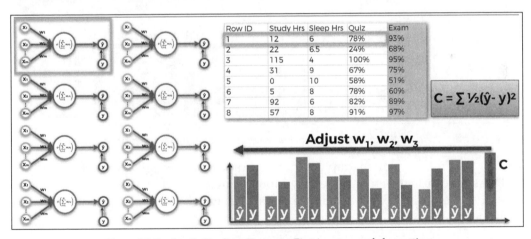

Figure 32: Stochastic Gradient Descent – First input row of observation

Second input row of observation:

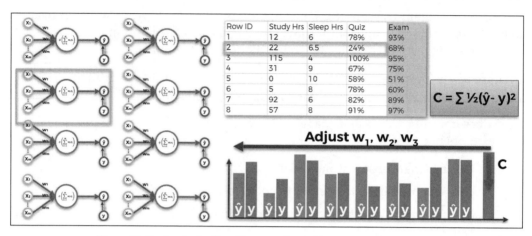

Figure 33: Stochastic Gradient Descent – Second input row of observation

Third input row of observation:

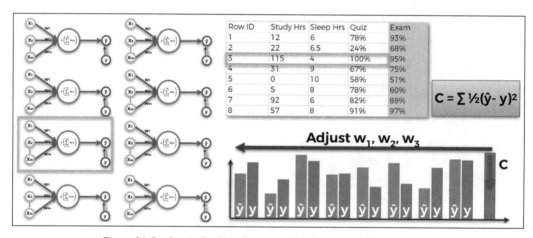

Figure 34: Stochastic Gradient Descent – Third input row of observation

Each of the preceding three graphics is an example of one weight's update run by SGD. As we can see, each time we only input a single row of observation from our dataset to the neural network, then we update the weights accordingly and proceed to the next input row of observation.

At first glance, SGD seems slower, because we input each row separately. In reality, it's much faster, because we don't have to load the whole dataset in the memory, nor wait for the whole dataset to pass through the model updating the weights.

To finish this section, let's recap the difference between batch gradient descent and SGD with the following graphic:

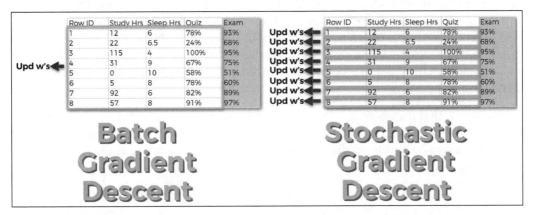

Figure 35: Batch Gradient Descent vs. Stochastic Gradient Descent

Now we can consider a middle-ground approach; mini-batch gradient descent.

Mini-batch gradient descent

Mini-batch gradient descent uses the best from both worlds, combining batch gradient descent with SGD. This is done by feeding the artificial neural network with small batches of data, instead of feeding single input rows of observations one by one or the whole dataset at once.

This approach is faster than classic SGD, and still prevents you from getting stuck in a local minimum. Mini-batch gradient descent also helps if you don't have enough computing resources to load the whole dataset in the memory, or enough processing power to get the full benefit of SGD.

That's all for neural networks! Now you're ready to combine your knowledge of neural networks with your knowledge of Q-learning.

Deep Q-learning

You've toured the foundations of deep learning, and you already know Q-learning; since deep Q-learning consists of combining Q-learning and deep learning, you're ready to get an intuitive grasp of deep Q-learning and crush it.

Before we start, try to guess some of how this is going to work. I would like you to take a moment and think about how you could integrate Q-learning into an ANN.

First things first, you might have guessed what the inputs and outputs of the neural network are going to be. The input of the artificial neural network is of course going to be the input state, which could be a 1-dimensional vector encoding what is happening in the environment, or an image (like the ones seen by a self-driving car). And the output is going to be the set of Q-values for each action, meaning it is going to be a 1-dimensional vector of several Q-values, one for each action that can be performed. Then, just like before, the AI takes the action that has the maximum Q-value, and performs it.

Very simply, that means that instead of predicting the Q-values through iterative updates with the Bellman equation (simple Q-learning), we'll predict them with an ANN that takes as inputs the input states, and returns as output the Q-values of the different actions.

That raises the question: it's good that we know what to predict, but what are going to be the targets (actual values) of these predictions when we are training the AI? As a reminder, the target is the actual value, or what you want your prediction to be ideally: the closer your prediction is to the target, the more it is correct. That's why we compute the loss error C between the prediction and the target, in order to reduce it through back-propagation with stochastic or mini-batch gradient descent.

When we were doing simple property price prediction, it was obvious what the targets were. They were simply the prices in the dataset that were available to us. But what about the targets of Q-values when you are training a self-driving car, for example? It's not that obvious, even though it is an explicit function of the Q-values and the reward.

The answer is a fundamental formula in deep Q-learning. The target of an input state s_t is:

$$R(s_t, a_t) + \gamma \max_a \left(Q(s_{t+1}, a) \right)$$

where $R(s_t, a_t)$ is the last reward obtained and γ is the discount factor, as seen previously.

Do you recognize the formula of the target? If you remember Q-learning, you should have no problem answering this question.

It's in the temporal difference, of course! Remember, the temporal difference is defined by:

$$TD_t\left(s_t,a_t\right)=R\left(s_t,a_t\right)+\gamma\max_a\left(Q\left(s_{t+1},a\right)\right)-Q\left(s_t,a_t\right)$$

So now it's obvious. The target is simply the first element at the left of the temporal difference:

$$\text{Target}=R\left(s_t,a_t\right)+\gamma\max_a\left(Q\left(s_{t+1},a\right)\right)$$

so that we get:

$$TD_t\left(s_t,a_t\right)=\text{Target}-Q\left(s_t,a_t\right)=\text{Target}-\text{Prediction}$$

Note that at the beginning, the Q-values are null, so the target is simply the reward.

There's one more piece to the puzzle before we can say that we really understand deep Q-learning; the Softmax method.

The Softmax method

This is the missing piece before we're ready to assemble everything for deep Q-learning. The Softmax method is the way we're going to select the action to perform after predicting the Q-values. In Q-learning, that was simple; the action performed was the one with the highest Q-value. That was the argmax method. In deep Q-learning, things are different. The problems are usually more complex, and so, in order to find an optimal solution, we must go through a process called **Exploration**.

Exploration consists of the following: instead of performing the action that has the maximum Q-value (which is called Exploitation), we're going to give each action a probability proportional to its Q-value, such that the higher the Q-value, the higher the probability. This creates, exactly, a distribution of the performable actions. Then finally, the action performed will be selected as a random draw from that distribution. Let me explain with an example.

Let's imagine we are building a self-driving car (we actually will, in *Chapter 10, AI for Autonomous Vehicles - Build a Self-Driving Car*). Let's say that the possible actions to perform are simple: move forward, turn left or turn right.

Then, at a specific time, let's say that our AI predicts the following Q-values:

Move Forward	Turn Left	Turn Right
24	38	11

The way we can create the distribution of probabilities we need is by dividing each Q-value by the sum of the three Q-values, which results each time in the probability of a particular action. Let's perform those sums:

$$\text{Probability of Moving Forward} = \frac{24}{24 + 38 + 11} = 33\%$$

$$\text{Probability of Turning Left} = \frac{38}{24 + 38 + 11} = 52\%$$

$$\text{Probability of Turning Right} = \frac{11}{24 + 38 + 11} = 15\%$$

Perfect – the probabilities sum to 1 and they are proportional to the Q-values. That gives us a distribution of the actions. To perform an action, the Softmax method takes a random draw from this distribution, such that:

- The action of Moving Forward has a 33% chance of being selected.
- The action of Turning Left has a 52% chance of being selected.
- The action of Turning Right has a 15% chance of being selected.

Can you feel the difference between Softmax and argmax, and do you understand why it is called Exploration instead of Exploitation? With argmax, the action *Turn Left* would be the one performed with absolute certainty. That's Exploitation. But with Softmax, even though the action *Turn Left* is the one with the highest chance of being selected, there's still a chance that the other actions might be selected.

Now, of course, the question is: why do we want to do that? It's simply because we want to explore the other actions, in case they lead to transitions resulting in higher rewards than we would obtain with pure exploitation. That often happens with complex problems, which are the ones for which deep Q-learning is used to find a solution. deep Q-learning finds that solution thanks to its advanced model, but also through exploration of the actions. This is a technique in AI called Policy Exploration.

As before, the next step is a step back. We're going to recap how deep Q-learning works.

Deep Q-learning recap

Deep Q-learning consists of combining Q-learning with an ANN.

Inputs are encoded vectors, each one defining a state of the environment. These inputs go into an ANN, where the output contains the predicted Q-values for each action.

More precisely, if there are n possible actions the AI could take, the output of the artificial neural network is a 1D vector comprised of n elements, each one corresponding to the Q-values of each action that could be performed in the current state. Then, the action performed is chosen via the Softmax method.

Hence, in each state s_t:

1. The prediction is the Q-value $Q(s_t, a_t)$, where a_t is performed by the Softmax method.
2. The target is $R(s_t, a_t) + \gamma \max_a (Q(s_{t+1}, a))$.
3. The loss error between the prediction and the target is the square of the temporal difference:

$$\text{Loss} = \frac{1}{2} \left(R(s_t, a_t) + \gamma \max_a (Q(s_{t+1}, a)) - Q(s_t, a_t) \right)^2 = \frac{1}{2} TD_t (s_t, a_t)^2$$

This loss error is back-propagated into the neural network, and the weights are updated according to how much they contributed to the error, through stochastic or mini-batch gradient descent.

Experience replay

You might noticed that so far we have only considered transitions from one state s_t to the next state s_{t+1}. The problem with this is that s_t is most of the time very correlated with s_{t+1}; therefore, the neural network is not learning much.

This could be improved if, instead of only considering the last transition each time, we considered the last m transitions, where m is a large number. This set of the last m transitions is what is called the experience replay memory, or simply memory. From this memory we sample some random transitions into small batches. Then we train the neural network with these batches to then update the weights through mini-batch gradient descent.

The whole deep Q-learning algorithm

Let's summarize the different steps of the whole deep Q-learning process.

Initialization:

1. Initialize the memory of the experience replay to an empty list M.
2. Choose a maximum size for the memory.

At each time t, we repeat the following process, until the end of the epoch:

1. Predict the Q-values of the current state s_t.
2. Perform the action selected by the Softmax method:

$$a_t = \underset{a}{\text{Softmax}}\left\{Q\left(s_t, a\right)\right\}$$

3. Get the reward $R\left(s_t, a_t\right)$.
4. Reach the next state s_{t+1}.
5. Append the transition $\left(s_t, a_t, r_t, s_{t+1}\right)$ to the memory M.
6. Take a random batch $B \subset M$ of transitions. For all the transitions $\left(s_{t_B}, a_{t_B}, r_{t_B}, s_{t_B+1}\right)$ of the random batch B:

    ○ Get the predictions: $Q\left(s_{t_B}, a_{t_B}\right)$

    ○ Get the targets: $R\left(s_{t_B}, a_{t_B}\right) + \gamma \max_a \left(Q\left(s_{t_B+1}, a\right)\right)$

    ○ Compute the loss between the predictions and the targets, over the whole batch B:

$$\text{Loss} = \frac{1}{2}\sum_B \left(R\left(s_{t_B}, a_{t_B}\right) + \gamma \max_a \left(Q\left(s_{t_B+1}, a\right)\right) - Q\left(s_{t_B}, a_{t_B}\right)\right)^2 = \frac{1}{2}\sum_B TD_{t_B}\left(s_{t_B}, a_{t_B}\right)^2$$

    ○ Back-propagate this loss error back into the neural network, and through stochastic gradient descent, update the weights according to how much they contributed to the loss error.

You've just unlocked the full deep Q-learning process! That means that you are now able to build powerful real-world AI applications in many fields. Here's a tour of some of the applications where deep Q-learning can create significant added value:

1. **Energy**: It was a deep Q-learning model that the DeepMind AI used to reduce Google's Data Center cooling bill by 40%. Also, deep Q-learning can optimize the functioning of smart grids; in other words, it can make smart grids even smarter.

2. **Transport**: Deep Q-learning can optimize traffic light control in order to reduce traffic.

3. **Autonomous Vehicles**: Deep Q-learning can be used to build self-driving cars, which we will illustrate in the next chapter of this book.

4. **Robotics**: Today, many advanced robots are built with deep Q-learning.

5. **And much more**: Chemistry, recommender systems, advertising, and many more—even video games, as you'll discover in *Chapter 13, AI for Games – Become the Master at Snake*, when you use deep convolutional Q-learning to train an AI to play Snake.

Summary

You learned a lot in this chapter; we first discussed ANNs. ANNs are built from neurons put in multiple layers. Each neuron from one layer is connected to every neuron from the previous layer, and every layer has its own activation function—a function that decides how much each output signal should be blocked.

The step in which an ANN works out the prediction is called forward-propagation and the step in which it learns is called back-propagation. There are three main types of back-propagation: batch gradient descent, stochastic gradient descent, and the best one, mini-batch gradient descent, which mixes the advantages of both previous methods.

The last thing we talked about in this chapter was deep Q-learning. This method uses Neural Networks to predict the Q-Values of taking certain actions. We also mentioned the experience replay memory, which stores a huge chunk of experience for our AI.

In the next chapter, you'll put all of this into practice by coding your very own self-driving car.

10

AI for Autonomous Vehicles – Build a Self-Driving Car

I'm really pumped up for you to start this new chapter. It's probably the most challenging, and most fun, adventure we'll have in this book. You're literally about to build a self-driving car from scratch, on a 2D map, using the powerful deep Q-learning model. I think that's incredibly exciting!

Think fast; what's our first step?

If you answered "building the environment," you're absolutely right. I hope that's getting so familiar to you that you answered before I even finished the question. Let's start by building an environment in which a car can learn how to drive by itself.

Building the environment

This time, we have much more to define than just the states, actions, and rewards. Building a self-driving car is a seriously complex problem. Now, I'm not going to ask you to go to your garage and turn yourself into a hybrid AI mechanic; you're simply going to build a virtual self-driving car that moves around a 2D map.

You'll build this 2D map inside a Kivy web app. Kivy is a free and open source Python framework, used for the development of applications like games, or really any kind of mobile app. Check out the website here: `https://kivy.org/#home`.

The whole environment for this project is built with Kivy, from start to finish. The development of the map and the virtual car has nothing to do with AI, so we won't go line by line through the code that implements it.

However, I am going to describe the features of the map. For those of you curious to know about exactly how the map is built, I've provided a fully commented Python file in the GitHub named `map_commented.py` that builds the environment from scratch with a full explanation.

Before we look at all the features, let's have a look at this map with the little virtual car inside:

Figure 1: The map

The first thing you'll notice is a black screen, which is the Kivy user interface. You build your games or apps inside this interface. As you might guess, it's actually the container of the whole environment.

You can see something weird inside, a white rectangle with three colored dots in front of it. Well, that's the car! My apologies for not being a better artist, but it's important to keep things simple. The white little rectangle is the shape of the car, and the three little dots are the sensors of the car. Why do we need sensors? Because on this map, we will have the option to build roads, delimited by sand, which the car will have to avoid going through.

To put some sand on the map, simply keep pressing left with your mouse and draw whatever you want. It doesn't have to just be roads; you can add some obstacles as well. In any case, the car will have to avoid going through the sand.

If you remember that everything works from the rewards, I'm sure you already know how to make that happen; it's by penalizing the self-driving car with a bad reward when it goes onto the sand. We'll take care of that later. In the meantime, let's have a look at one of my nice drawings of roads with sand:

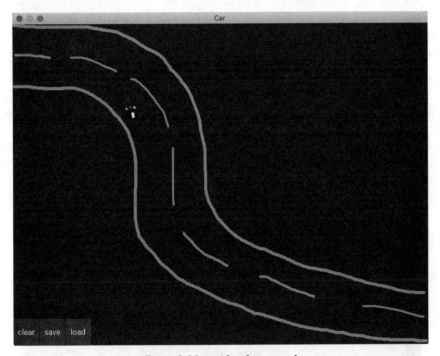

Figure 2: Map with a drawn road

The sensors are there to detect the sand, so the car can avoid it. The blue sensor covers an area at the left of the car, the red sensor covers an area at the front of the car, and the yellow sensor covers an area at the right of the car.

Finally, there are three buttons to click on at the bottom left corner of the screen, which are:

clear: Removes all the sand drawn on the map

save: Saves the weights (parameters) of the AI

load: Loads the last saved weights

Now we've had a look at our little map, let's move on to defining our goals.

Defining the goal

We understand that our goal is to build a self-driving car. Good. But how are we going to formalize that goal, in terms of AI and reinforcement learning? Your intuition should hopefully make you think about the rewards we're going to set. I agree — we're going to give a high reward to our car if it manages to self-drive. But how can we tell that it's managing to self-drive?

We've got plenty of ways to evaluate this. For example, we could simply draw some obstacles on the map, and train our self-driving car to move around the map without hitting the obstacles. That's a simple challenge, but we could try something a little more fun. Remember the road I drew earlier? How about we train our car to go from the upper left corner of the map, to the bottom right corner, through any road we build between these two spots? That's a real challenge, and that's what we'll do. Let's imagine that the map is a city, where the upper left corner is the Airport, and the bottom right corner is Downtown:

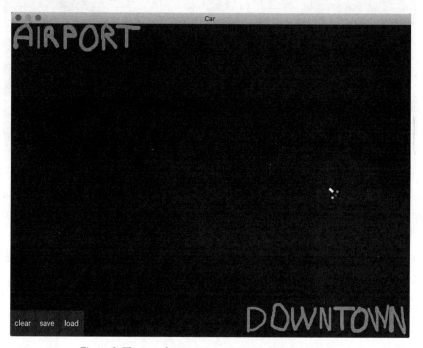

Figure 3: The two destinations – Airport and Downtown

Now we can clearly formulate a goal; to train the self-driving car to make round trips between the Airport and Downtown. As soon as it reaches the Airport, it will then have to go to Downtown, and as soon as it reaches Downtown, it will then have to go the Airport. More than that, it should be able to make these round trips along any road connecting these two locations. It should also be able to cope with any obstacles along that road it has to avoid. Here is an example of another, more challenging road:

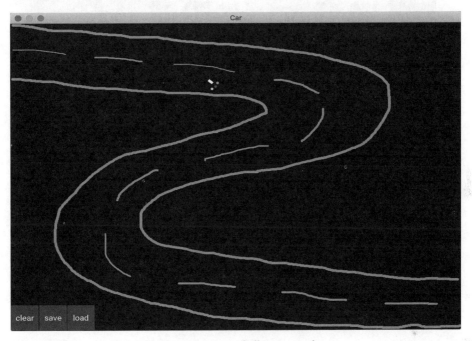

Figure 4: A more challenging road

If you think that road look too easy, here's a more challenging example; this time with not only a more difficult road but also many obstacles:

Figure 5: An even more challenging road

As a final example, I want to share this last map, designed by one of my students, which could belong in the movie *Inception*:

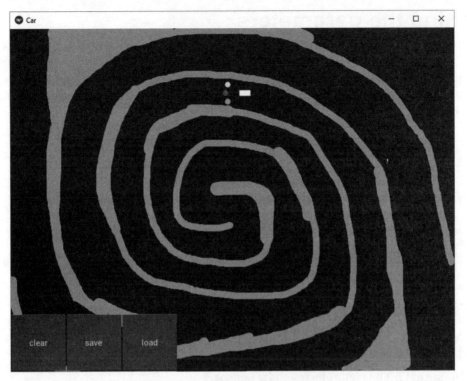

Figure 6: The most challenging road ever

If you look closely, it's still a path that goes from Airport to Downtown and vice versa, just much more challenging. The AI we create will be able to cope with any of these maps.

I hope you find that as exciting as I do! Keep that level of energy up, because we have quite a lot of work to do.

Setting the parameters

Before you define the input states, the output actions, and the rewards, you must set all of the parameters of the map and the car that will be part of your environment. The inputs, outputs, and rewards are all functions of these parameters. Let's list them all, using the same names as in the code, so that you can easily understand the file map.py:

1. **angle**: The angle between the *x*-axis of the map and the axis of the car

2. **rotation**: The last rotation made by the car (we will see later that when playing an action, the car makes a rotation)

3. **pos = (self.car.x, self.car.y)**: The position of the car (self.car.x is the *x*-coordinate of the car, self.car.y is the *y*-coordinate of the car)

4. **velocity = (velocity_x, velocity_y)**: The velocity vector of the car

5. **sensor1 = (sensor1_x, sensor1_y)**: The position of the first sensor

6. **sensor2 = (sensor2_x, sensor2_y)**: The position of the second sensor

7. **sensor3 = (sensor3_x, sensor3_y)**: The position of the third sensor

8. **signal1**: The signal received by sensor 1

9. **signal2**: The signal received by sensor 2

10. **signal3**: The signal received by sensor 3

Now let's slow down; we've got to define how these signals are computed. The signals are a measure of the density of sand around their sensor. How are you going to compute that density? You start by introducing a new variable, called sand, which you initialize as an array that has as many cells as our graphic interface has pixels. Simply put, the sand array is the black map itself and the pixels are the cells of the array. Then, each cell of the sand array will get a 1 if there is sand, and a 0 if there is not.

For example, here the sand array has only 1s in its first few rows, and the rest is all 0s:

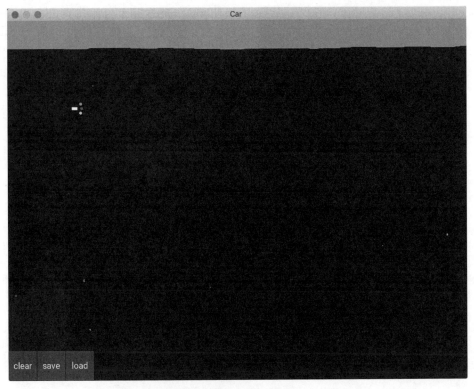

Figure 7: The map with only sand in the first rows

I know the border is a little wobbly—like I said, I'm no great artist—and that just means those rows of the sand array would have 1s where the sand is and 0s where there's no sand.

Now that you have this sand array it's very easy to compute the density of sand around each sensor. You surround your sensor by a square of 20 by 20 cells (which the sensor reads from the sand array), then you count the number of ones in these cells, and finally you divide that number by the total number of cells in that square, that is, 20 x 20 = 400 cells.

Since the `sand` array only contains 1s (where there's sand) and 0s (where there's no sand), we can very easily count the number of 1s by simply summing the cells of the `sand` array in this 20 by 20 square. That gives us exactly the density of sand around each sensor, and that's what's computed at lines 81, 82, and 83 in the `map.py` file:

```
81          self.signal1 = int(np.sum(sand[int(self.sensor1_x)-
      10:int(self.sensor1_x)+10, int(self.sensor1_y)-10:int(self.
      sensor1_y)+10]))/400.

82          self.signal2 = int(np.sum(sand[int(self.sensor2_x)-
      10:int(self.sensor2_x)+10, int(self.sensor2_y)-10:int(self.
      sensor2_y)+10]))/400.

83          self.signal3 = int(np.sum(sand[int(self.sensor3_x)-
      10:int(self.sensor3_x)+10, int(self.sensor3_y)-10:int(self.
      sensor3_y)+10]))/400.
```

Now that we've covered how the signals are computed, let's continue with the rest of the parameters. The last parameters, which I've highlighted in the list below, are important because they're the last pieces that we need to reveal the final input state vector. Here they are:

1. **goal_x**: The *x*-coordinate of the goal (which can either be the Airport or Downtown)

2. **goal_y**: The *y*-coordinate of the goal (which can either be the Airport or Downtown)

3. **xx = (goal_x - self.car.x)**: The difference of *x*-coordinates between the goal and the car

4. **yy = (goal_y - self.car.y)**: The difference of *y*-coordinates between the goal and the car

5. **orientation**: The angle that measures the direction of the car with respect to the goal

Let's slow down again for a moment. We need to know how orientation is computed; it's the angle between the axis of the car (the `velocity` vector from our first list of parameters) and the axis that joins the goal and the center of the car. The goal has the coordinates (`goal_x, goal_y`) and the center of the car has the coordinates (`self.car.x, self.car.y`). For example, if the car is heading perfectly toward the goal, then orientation = 0°. If you're curious as to how we can compute the angle between the two axes in Python, here's the code that gets the `orientation` (lines 126, 127, and 128 in the `map.py` file):

```
126          xx = goal_x - self.car.x
127          yy = goal_y - self.car.y
128          orientation = Vector(*self.car.velocity).
      angle((xx,yy))/180.
```

Good news—we're finally ready to define the main pillars of the environment. I'm talking, of course, about the input states, the actions, and the rewards.

Before I define them, try to guess what they're going to be. Check out all the preceding parameters again, and remember the goal: making round trips between two locations, the Airport and Downtown, while avoiding any obstacles along the road. The solution's in the next section.

The input states

What do you think the input states are? You might have answered "the position of the car." In that case, the input state would be a vector of two elements, the coordinates of the car: `self.car.x` and `self.car.y`.

That's a good start. From the intuition and foundation techniques of deep Q-learning you learned in *Chapter 9, Going Pro with Artificial Brains – Deep Q-Learning*, you know that when you're doing deep Q-learning, the input state doesn't have to be a single element as in Q-learning. In fact, in deep Q-learning the input state can be a vector of many elements, allowing you to supply many sources of information to your AI to help it predict smart actions to play.

 The input state can even be bigger than a simple vector: it can be an image! In that case, the AI model is called **deep convolutional Q-learning**. It's the same as deep Q-learning, except that you add a convolutional neural network at the entrance of the neural network that allows your AI (machine) to visualize images. We'll cover this technique in *Chapter 12, Deep Convolution Q-Learning*.

We can do better than just supplying the car position coordinates. They tell us where the self-driving car is located, but there's another parameter that's better, simpler, and more directly related to the goal. I'm talking about the `orientation` variable. The orientation is a single input that directly tells us if we are pointed in the right direction, toward the goal. If we have that orientation, we don't need the car position coordinates at all to navigate toward the goal; we can just change the orientation by a certain angle to point the car more in the direction of the goal. The actions that the AI performs will be what changes that orientation. We'll discuss those in the next section.

We have the first element of our input state: the orientation.

But that's not enough. Remember that we also have another goal, or, should I say, constraint. Our car needs to stay on the road and avoid any obstacles along that road.

In the input state, we need information telling the AI whether it is about to move off the road or hit an obstacle. Try and work it out for yourself—do we have a way to get this information?

The solution is the sensors. Remember that our car has three sensors giving us signals about how much sand is around them. The blue sensor tells us if there's any sand at the left of the car, the red sensor tells us if there is any sand in front of the car, and the yellow sensor tells us if there is any sand at the right of the car. The signals of these sensors are already coded into three variables: `signal1`, `signal2`, and `signal3`. These signals will tell the AI if it's about to hit some obstacle or about to get out of the road, since the road is delimited by sand.

That's the rest of the information you need for your input state. With these four elements, `signal1`, `signal2`, `signal3`, and `orientation`, you have everything you need to be able to drive from one location to another, while staying on the road, and without hitting any obstacles.

In conclusion, here's what the input state is going to be at each time:

Input state = (`orientation`, `signal1`, `signal2`, `signal3`)

And that's exactly what's coded at line 129 in the `map.py` file:

```
129              state = [orientation, self.car.signal1, self.car.
         signal2, self.car.signal3]
```

`state` is the variable name given to the input state.

Don't worry too much about the code syntax difference between `signal`, `self.signal`, and `self.car.signal`; they're all the same. The reason we use these different variables is because the AI is coded with classes (as in **Object Oriented Programming (OOP)**), which allows us to create several self-driving cars on the same map.

If you do want to have several self-driving cars on your map, for example, if you want them racing, then you can distinguish the cars better thanks to `self.car.signal`. For example, if you have two cars, you can name the two objects `car1` and `car2` so that you can distinguish the first sensor signals of the two cars, by using `self.car1.signal1` and `self.car2.signal1`. In this chapter, we just have one car, so whether we use `signal1`, `car.signal1` or `self.car.signal1`, we get the same thing.

We've covered the input state; now let's tackle the actions.

The output actions

I've already briefly mentioned or suggested what the actions are going to be. Given our input state, it's easy to guess. Naturally, since you're building a self-driving car, you might think that the actions should be: move forward, turn left, or turn right. You'd be absolutely right! That's exactly what the actions are going to be.

Not only is this intuitive, but it aligns extremely well with our choice of input states. They contain the `orientation` variable that tells us if we're aimed in the right direction toward the goal. Simply put, if the `orientation` input tells us our car is pointed in the right direction, we perform the action of moving forward. If the `orientation` input tells us that the goal is on the right of our car, we perform the action of turning right. Finally, if the `orientation` tells us that the goal is on the left of our car, we perform the action of turning left.

At the same time, if any of the signals spot some sand around the car, the car will turn left or right to avoid it. The three possible actions of move forward, turn left, and turn right make logical sense with the goal, constraint, and input states we have, and we can define them as the three following rotations:

rotations = [turn 0° (that is, move forward), turn 20° to the left, turn 20° to the right]

The choice of 20° is quite arbitrary. You could very well choose 10°, 30°, or 40°. I'd avoid more than 40°, because then your car would have twitchy, fidgety movements, and wouldn't look like a smoothly moving car.

However, the actions the ANN outputs will not be 0°, 20°, and -20°; they will be 0, 1 and 2.

actions = [0, 1, 2]

It's always better to use simple categories like those when you're dealing with the output of an artificial neural network. Since 0, 1, and 2 will be the actions the AI returns, how do you think we end up with the rotations?

You'll use a simple mapping, called `action2rotation` in our code, which maps the actions 0, 1, 2 to the respective rotations of 0°, 20°, -20°. This is exactly what's coded on lines 34 and 131 of the `map.py` file:

```
34      action2rotation = [0,20,-20]
```

```
131          rotation = action2rotation[action]
```

Now, let's move on to the rewards. This one's going to be fun, because this is where you decide how you want to reward or punish your car. Try to figure out how by yourself first, and then take a look at the solution in the following section.

The rewards

To define the system of rewards, we have to answer the following questions:

- In which cases do we give the AI a good reward? How good for each case?
- In which cases do we give the AI a bad reward? How bad for each case?

To answer these questions, we must simply remember what the goal and constraints are:

- The goal is to make round trips between the Airport and Downtown.
- The constraints are to stay on the road and avoid obstacles if any. In other words, the constraint is to stay away from the sand.

Hence, based on this goal and constraints, the answers to our preceding questions are:

1. We give the AI a good reward when it gets closer to the destination.
2. We give the AI a bad reward when it gets further away from the destination.
3. We give the AI a bad reward if it's about to drive onto some sand.

That's it! That should work, because these good and bad rewards have a direct effect on the goal and constraints.

To answer the second part of each question, how good and how bad the reward should be for each case, we'll play the tough card; it's often more effective. The tough card consists of punishing the car more when it makes mistakes than we reward it when it does well. In other words, the bad reward is going to be stronger than the good reward.

This works well in reinforcement learning, but that doesn't mean you should do the same with your dog or your kids. When you're dealing with a biological system, the other way around (high good reward and small bad reward) is a much more effective way to train or educate. Just food for thought.

On that note, here are the rewards we'll give in each case:

1. The AI gets a bad reward of -1 if it drives onto some sand. Nasty!
2. The AI gets a bad reward of -0.2 if it moves away from the destination.
3. The AI gets a good reward of 0.1 if it moves closer to the destination.

The reason we attribute the worst reward (-1) to the case when the car drives onto some sand makes sense. Driving onto sand is what we absolutely want to avoid. The sand on the map represents obstacles in real life; in real life, you would train your self-driving car not to hit any obstacle, so as to avoid any accident. To do so, we penalize the AI with a highly bad reward when it does hit an obstacle during its training.

How's that translated that into code? That's easy; you just take your sand array and check if the car has just moved onto a cell that contains a 1. If it does, that means the car has moved onto some sand and must therefore get a bad reward of -1. That's exactly what's coded here at lines 138, 139, and 140 of the map.py file (including an update of the car velocity vector, which not only updates the speed by slowing the car down to 1, but also updates the direction of the car by a certain angle, self.car. angle):

```
138          if sand[int(self.car.x),int(self.car.y)] > 0:
139               self.car.velocity = Vector(1, 0).rotate(self.car.
      angle)
140               reward = -1
```

Then for the other reward attributions, you just have to complete the if condition preceding with an else, which will say what happens in the case where the car has not driven onto some sand.

In that case, you start a new if and else condition, saying that if the car has moved away from the destination, you give it a bad reward of -0.2, and, if the car has moved closer to the destination, you give it a good reward of 0.1. The way you measure if the car is getting away from or closer to the goal is by comparing two distances put into two separate variables: last_distance, which is the previous distance between the car and the destination at time *t*-1, and distance, which is the current distance between the car and the destination at time *t*. If you put all that together, you get the following code, which completes the preceding lines of code:

```
138          if sand[int(self.car.x),int(self.car.y)] > 0:
139               self.car.velocity = Vector(1, 0).rotate(self.car.
      angle)
140               reward = -1
141          else:
142               self.car.velocity = Vector(6, 0).rotate(self.car.
      angle)
143               reward = -0.2
144               if distance < last_distance:
145                    reward = 0.1
```

To keep the car from trying to veer off the map, lines 147 to 158 of the `map.py` file punish the AI with a bad reward of -1 if the self-driving car gets within 10 pixels of any of the map's 4 borders of the map. Finally, lines 160 to 162 of the `map.py` file update the goal, switching it from the Airport to Downtown, or vice versa, anytime the car gets within 100 pixels of the current goal.

AI solution refresher

Let's refresh our memory by reminding ourselves of the steps of the deep Q-learning process, while adapting them to our self-driving car application.

Initialization:

1. The memory of the experience replay is initialized to an empty list, called **memory** in the code.

2. The maximum size of the memory is set, called **capacity** in the code.

At each time *t*, the AI repeats the following process, until the end of the epoch:

1. The AI predicts the Q-values of the current state s_t. Therefore, since three actions can be played (0 <-> 0°, 1 <-> 20°, or 2 <-> -20°), it gets three predicted Q-values.

2. The AI performs an action selected by the Softmax method (see *Chapter 5, Your First AI Model – Beware the Bandits!*):

$$a_t = \operatorname*{Softmax}_a \{Q(s_t, a)\}$$

3. The AI receives a reward $R(s_t, a_t)$, which is one of -1, -0.2 or +0.1.

4. The AI reaches the next state s_{t+1}, which is composed of the next three signals from the three sensors, plus the orientation of the car.

5. The AI appends the transition (s_t, a_t, r_t, s_{t+1}) to the memory.

6. The AI takes a random batch $B \subset M$ of transitions. For all the transitions $(s_{t_B}, a_{t_B}, r_{t_B}, s_{t_B+1})$ of the random batch B:

    ○ The AI gets the predictions: $Q(s_{t_B}, a_{t_B})$

    ○ The AI gets the targets: $R(s_{t_B}, a_{t_B}) + \gamma \max_a (Q(s_{t_B+1}, a))$

    ○ The AI computes the loss between the predictions and the targets over the whole batch B:

$$Loss = \frac{1}{2}\sum_{B}\left(R\left(s_{t_B},a_{t_B}\right)+\gamma\max_{a}\left(Q\left(s_{t_B+1},a\right)\right)-Q\left(s_{t_B},a_{t_B}\right)\right)^2 = \frac{1}{2}\sum_{B}TD_{t_B}\left(s_{t_B},a_{t_B}\right)^2$$

○ Finally, the AI backpropagates this loss error into the neural network, and through stochastic gradient descent updates the weights according to how much they contributed to the loss error.

Implementation

Now it's time for the implementation! The first thing you need is a professional toolkit, because you're not going to build an artificial brain with simple Python libraries. What you need is an advanced framework, which allows fast computation for the training of neural networks.

Today, the best frameworks to build and train AIs are **TensorFlow** (by Google) and **PyTorch** (by Facebook). How should you choose between the two? They're both great to work with and equally powerful. They both have dynamic graphs, which allow the fast computation of the gradients of complex functions needed to train the model during backpropagation with mini-batch gradient descent. Really, it doesn't matter which framework you choose; both work very well for our self-driving car. As far as I'm concerned, I have slightly more experience with PyTorch, so I'm going to opt for PyTorch and that's how the example in this chapter will continue to play out.

To take a step back, our self-driving car implementation is composed of three Python files:

1. `car.kv`, which contains the Kivy objects (rectangle shape of the car and the three sensors)

2. `map.py`, which builds the environment (map, car, input states, output actions, rewards)

3. `deep_q_learning.py`, which builds and trains the AI through deep Q-learning

We've already covered the major elements of `map.py`, and now we're about to tackle `deep_q_learning.py`, where you'll not only build an artificial neural network, but also implement the deep Q-learning training process. Let's get started!

Step 1 – Importing the libraries

As usual, you start by importing the libraries and modules you need to build your AI. These include:

1. os: The operating system library, used to load the saved AI models.

2. random: Used to sample some random transitions from the memory for experience replay.

3. torch: The main library from PyTorch, which will be used to build our neural network with tensors, as opposed to simple matrices like numpy arrays. While a matrix is a 2-D array, a tensor can be a *n*-dimensional array, with more than just a single number in its cells. Here's a diagram so you can clearly understand the difference between a matrix and a tensor:

$$\begin{bmatrix} 1 & 2 \\ 3 & 4 \end{bmatrix} \qquad \begin{bmatrix} \begin{bmatrix} 1 & 2 \\ 3 & 4 \end{bmatrix} & \begin{bmatrix} 5 & 6 \\ 7 & 8 \end{bmatrix} \\ \begin{bmatrix} 9 & 10 \\ 11 & 12 \end{bmatrix} & \begin{bmatrix} 13 & 14 \\ 15 & 16 \end{bmatrix} \end{bmatrix}$$

Matrix   Tensor

4. torch.nn: The nn module from the torch library, used to build the fully connected layers in the artificial neural network of our AI.

5. torch.nn.functional: The functional sub-module from the nn module, used to call the activation functions (rectifier and Softmax), as well as the loss function for backpropagation.

6. torch.optim: The optim module from the torch library, used to call the Adam optimizer, which computes the gradients of the loss with respect to the weights and updates those weights in directions that reduce the loss.

7. torch.autograd: The autograd module from the torch library, used to call the Variable class, which associates each tensor and its gradient into the same variable.

That makes up your first code section:

```
1       # AI for Autonomous Vehicles - Build a Self-Driving Car
2
3       # Importing the libraries
4
```

```
5       import os
6       import random
7       import torch
8       import torch.nn as nn
9       import torch.nn.functional as F
10      import torch.optim as optim
11      from torch.autograd import Variable
```

Step 2 – Creating the architecture of the neural network

This code section is where you really become the architect of the brain in your AI. You're about to build the input layer, the fully connected layers, and the output layer, while choosing some activation functions that will forward-propagate the signal inside the brain.

First, you build this brain inside a class, which we are going to call `Network`.

What is a class? Let's explain that before we explain why you're using one. A class is an advanced structure in Python that contains the instructions of an object we want to build. Taking the example of your neural network (the object), these instructions include how many layers you want, how many neurons you want inside each layer, which activation function you choose, and so on. These parameters define your artificial brain and are all gathered in what we call the __init__() method, which is what we always start with when building a class. But that's not all—a class can also contain tools, called methods, which are functions that either perform some operations or return something. Your `Network` class will contain one method, which forward-propagates the signal inside the neural network and returns the predicted Q-values. Call this method `forward`.

Now, why use a class? That's because building a class allows you to create as many objects (also called instances) as you want, and easily switch from one to another by just changing the arguments of the class. For example, your `Network` class contains two arguments: `input_size` (the number of inputs) and `nb_actions` (the number of actions). If you ever want to build an AI with more inputs (besides the signals and the orientation) or more outputs (you could add an action that brakes the car), you'll do it in a flash thanks to the advanced structure of the class. It's super practical, and if you're not already familiar with classes you'll have to get familiar with them. Nearly all AI implementations are done with classes.

That was just a short technical aside to make sure I don't lose anybody on the way. Now let's build this class. As there are many important elements to explain in the code, and since you're probably new to PyTorch, I'll show you the code first and then explain it line by line from the `deep_q_learning.py` file:

```
13      # Creating the architecture of the Neural Network
14
15      class Network(nn.Module):
16
17          def __init__(self, input_size, nb_action):
18              super(Network, self).__init__()
19              self.input_size = input_size
20              self.nb_action = nb_action
21              self.fc1 = nn.Linear(input_size, 30)
22              self.fc2 = nn.Linear(30, nb_action)
23
24          def forward(self, state):
25              x = F.relu(self.fc1(state))
26              q_values = self.fc2(x)
27              return q_values
```

Line 15: You introduce the `Network` class. In the parenthesis of this class, you can see `nn.Module`. That means you're calling the `Module` class, which is an existing class taken from the `nn` module, in order to get all the properties and tools of the `Module` class and use them inside your `Network` class. This trick of calling another existing class inside a new class is called **inheritance**.

Line 17: You start with the __init__ () method, which defines all the parameters (number of inputs, number of outputs, and so on) of your artificial neural network. You can see three arguments: `self`, `input_size`, and `nb_action`. `self` refer to the object, that is, to the future instance of the class that will be created after the class is done. Any time you see `self` before a variable, and separated by a dot (like `self.variable`), that means the variable belongs to the object. That should clear up any mystery about `self`!

Then, `input_size` is the number of inputs in your input state vector (thus 4), and `nb_action` is the number of output actions (thus 3). What's important to understand is that the arguments (other than self) of the __init__ () method are the ones you will enter when creating the future object, which is the future artificial brain of your AI.

Line 18: You use the `super()` function to activate the inheritance (explained in Line 15), inside the `__init__()` method.

Line 19: Here you introduce the first object variable, `self.input_size`, set equal to the argument `input_size` (which will later be entered as 4, since the input state has 4 elements).

Line 20: You introduce the second object variable, `self.nb_action`, set equal to the argument `nb_action` (which will later be entered as 3, since there are three actions that can be performed).

Line 21: You introduce the third object variable, `self.fc1`, which is the first full connection between the input layer (composed of the input state) and the hidden layer. That first full connection is created as an object of the `nn.Linear` class, which takes two arguments: the first one is the number of elements in the left layer (the input layer), so `input_size` is the right argument to use, and the second one is the number of hidden neurons in the right layer (the hidden layer). Here, you choose to have 30 neurons, and therefore the second argument is 30. The choice of 30 is purely arbitrary, and the self-driving car could work well with any other numbers.

Line 22: You introduce the fourth object variable, `self.fc2`, which is the second full connection between the hidden layer (composed of 30 hidden neurons) and the output layer. It could have been a full connection with a new hidden layer, but your problem is not complex enough to need more than one hidden layer, so you'll just have one hidden layer in your artificial brain. Just like before, that second full connection is created as an object of the `nn.Linear` class, which takes two arguments: the first one is the number of elements in the left layer (the hidden layer), therefore 30, and the second one is the number of hidden neurons in the right layer (the output layer), therefore 3.

Line 24: You start building the first and only method of the class, the `forward` method, which will propagate the signal from the input layer to the output layer, after which it will return the predicted Q-values. This `forward` method takes two arguments: `self`, because you'll use the object variables inside the `forward` method, and `state`, the input state vector composed of four elements (orientation plus the three signals).

Line 25: You forward propagate the signal from the input layer to the hidden layer while activating the signal with a rectifier activation function, also called **ReLU** (**Rectified Linear Unit**). You do this in two steps. First, the forward propagation from the input layer to the hidden layer is done by calling the first full connection `self.fc1` with the input state vector `state` as input: `self.fc1(state)`.

That returns the hidden layer. And then we call the `relu` function with that hidden layer as input to break the linearity of the signal the following way:

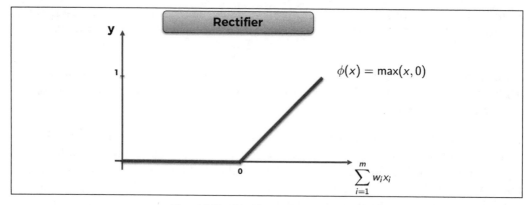

Figure 8: The Rectifier activation function

The purpose of the ReLU layer is to break linearity by creating non-linear operations along the fully connected layers. You'll want to have that, because you're trying to solve a nonlinear problem. Finally, `F.relu(self.fc1(state))` returns x, the hidden layer with a nonlinear signal.

Line 26: You forward-propagate the signal from the hidden layer to the output layer containing the Q-values. In the same way as the previous line, this is done by calling the second full connection `self.fc2` with the hidden layer x as input: `self.fc2(x)`. That returns the Q-values, which you name `q_values`. Here, no activation function is needed because you'll select the action to play with Softmax, later, in another class.

Line 27: Finally, here, the `forward` method returns the Q-values.

Let's have a look at what you've just created!

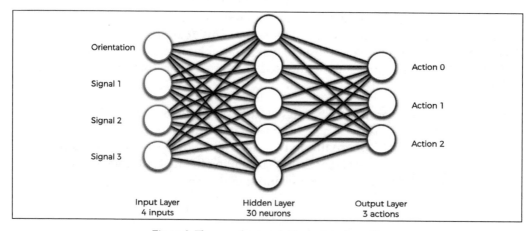

Figure 9: The neural network (the brain) of our AI

`self.fc1` are all the blue connection lines between the **Input Layer** and the **Hidden Layer**.

`self.fc2` are all the blue connection lines between the **Hidden Layer** and the **Output Layer**.

That should help you visualize the full connections better. Great job!

Step 3 – Implementing experience replay

Time for the next step! You'll now build another class, which builds the memory object for experience replay (as seen in *Chapter 5, Your First AI Model – Beware the Bandits!*). Call this class `ReplayMemory`. Let's have a look at the code first and then I'll explain everything line by line from the `deep_q_learning.py` file.

```
29    # Implementing Experience Replay
30
31    class ReplayMemory(object):
32
33        def __init__(self, capacity):
34            self.capacity = capacity
35            self.memory = []
36
37        def push(self, event):
38            self.memory.append(event)
39            if len(self.memory) > self.capacity:
40                del self.memory[0]
41
42        def sample(self, batch_size):
43            samples = zip(*random.sample(self.memory, batch_size))
44            return map(lambda x: Variable(torch.cat(x, 0)), samples)
```

Line 31: You introduce the `ReplayMemory` class. This time you don't need to inherit from any other class, so just input `object` in the parenthesis of the class.

Line 33: As always, you start with the `__init__`() method, which only takes two arguments: `self`, the object, and `capacity`, the maximum size of the memory.

Line 34: You introduce the first object variable, `self.capacity`, set equal to the argument `capacity`, which will be entered later when creating an object of the class.

Line 35: You introduce the second object variable, `self.memory`, initialized as an empty list.

Line 37: You start building the first tool of the class, the push method, which takes a transition as input and adds it to the memory. However, if adding that transition exceeds the memory's capacity, the push method also deletes the first element of the memory. The event argument you can see is the transition to be added.

Line 38: Using the append function, you add the transition to the memory.

Line 39: You start an if condition that checks if the length of the memory (meaning its number of transitions) is larger than the capacity.

Line 40: If that is indeed the case, you delete the first element of the memory.

Line 42: You start building the second tool of the class, the sample method, which samples some random transitions from the experience replay memory. It takes batch_size as input, which is the size of the batches of transitions with which you'll train your neural network.

Remember how it works: instead of forward-propagating single input states into the neural network and updating the weights after each transition resulting from the input state, you forward-propagate small batches of input states and update the weights after backpropagating the same whole batches of transitions with mini-batch gradient descent. That's different from stochastic gradient descent (weight update every single input) and batch gradient descent (weight update every batch of inputs) as explained in *Chapter 9, Going Pro with Artificial Brains – Deep Q-Learning*:

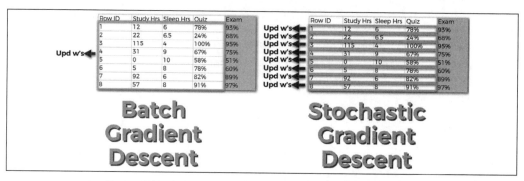

Figure 10: Batch gradient descent versus stochastic gradient descent

Line 43: You sample some random transitions from the memory and put them into a batch of size batch_size. For example, if batch_size = 100, you sample 100 random transitions from the memory. The sampling is done with the sample() function from the random library. Then, zip(*list) is used to regroup the states, actions, and rewards into separate batches of the same size (batch_size), in order to put the sampled transitions into the format expected by PyTorch (the Variable format, which comes next in Line 44).

This is probably a good time to take a step back. Let's see what Line 43 gives you:

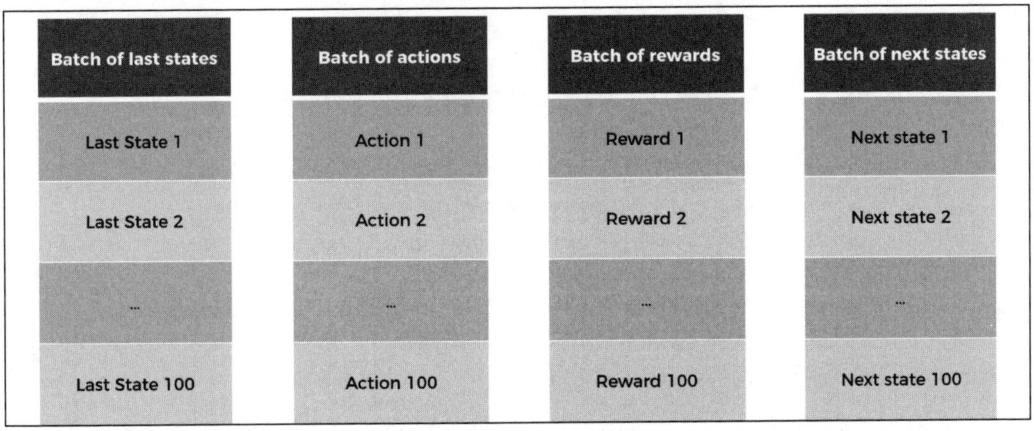

Figure 11: The batches of last states, actions, rewards, and next states

Line 44: Using the map() function, wrap each sample into a torch Variable object (as Variable() is actually a class), so that each tensor inside the samples is associated to a gradient. Simply put, you can see a torch Variable as an advanced structure that encompasses a tensor and a gradient.

This is the beauty of PyTorch. These torch Variables are all in a dynamic graph which allows fast computation of the gradient of complex functions. Those fast computations are required for the weight updates happening during backpropagation with mini-batch gradient descent. Inside the Variable class we see torch.cat(x,0). That's just a concatenation trick, along the vertical axis, to put the samples in the format expected by the Variable class.

The most important thing to remember is this: when training a neural network with PyTorch, we always work with torch Variables, as opposed to just tensors. You can find more details about this in the PyTorch documentation.

Step 4 – Implementing deep Q-learning

You've made it! You're finally about to start coding the whole deep Q-learning process. Again, you'll wrap all of it into a class, this time called Dqn, as in deep Q-network. This is your final run before the finish line. Let's smash this.

This time, the class is quite long so I'll show and explain the lines of code method by method from the deep_q_learning.py file. Here's the first one, the __init__() method:

```
46    # Implementing Deep Q-Learning
```

```
47
48     class Dqn(object):
49
50         def __init__(self, input_size, nb_action, gamma):
51             self.gamma = gamma
52             self.model = Network(input_size, nb_action)
53             self.memory = ReplayMemory(capacity = 100000)
54             self.optimizer = optim.Adam(params = self.model.
       parameters())
55             self.last_state = torch.Tensor(input_size).unsqueeze(0)
56             self.last_action = 0
57             self.last_reward = 0
```

Line 48: You introduce the Dqn class. You don't need to inherit from any other class so just input object in the parenthesis of the class.

Line 50: As always, you start with the __init__() method, which this time takes four arguments:

1. self: The object
2. input_size: The number of inputs in the input state vector (that is, 4)
3. nb_action: The number of actions (that is, 3)
4. gamma: The discount factor in the temporal difference formula

Line 51: You introduce the first object variable, self.gamma, set equal to the argument gamma (which will be entered later when you create an object of the Dqn class).

Line 52: You introduce the second object variable, self.model, an object of the Network class you built before. This object is your neural network; in other words, the brain of our AI. When creating this object, you input the two arguments of the __init__() method in the Network class, which are input_size and nb_action. You'll enter their real values (respectively 4 and 3) later, when creating an object of the Dqn class.

Line 53: You introduce the third object variable, self.memory, as an object of the ReplayMemory class you built before. This object is the experience replay memory. Since the __init__ method of the ReplayMemory class only expects one argument, the capacity, that's exactly what you input here as 100,000. In other words, you're creating a memory of size 100,000, which means that instead of remembering just the last transition, the AI will remember the last 100,000 transitions.

Line 54: You introduce the fourth object variable, `self.optimizer`, as an object of the `Adam` class, which is an existing class built in the `torch.optim` module. This object is the optimizer, which updates the weights through mini-batch gradient descent during backpropagation. In the arguments, keep most of the default parameter values (you can check them in the PyTorch documentation) and only enter the model parameters (the `params` argument), which you access with `self.model.parameters`, one of the attributes of the `nn.Module` class from which the `Network` class inherits.

Line 55: You introduce the fifth object variable, `self.last_state`, which will be the last state in each (last state, action, reward, next state) transition. This last state is initialized as an object of the `Tensor` class from the torch library, into which you only have to enter the `input_size` argument. Then `.unsqueeze(0)` is used to create an additional dimension at index 0, which will correspond to the batch. This allows us to do something like this, matching each last state to the appropriate batch:

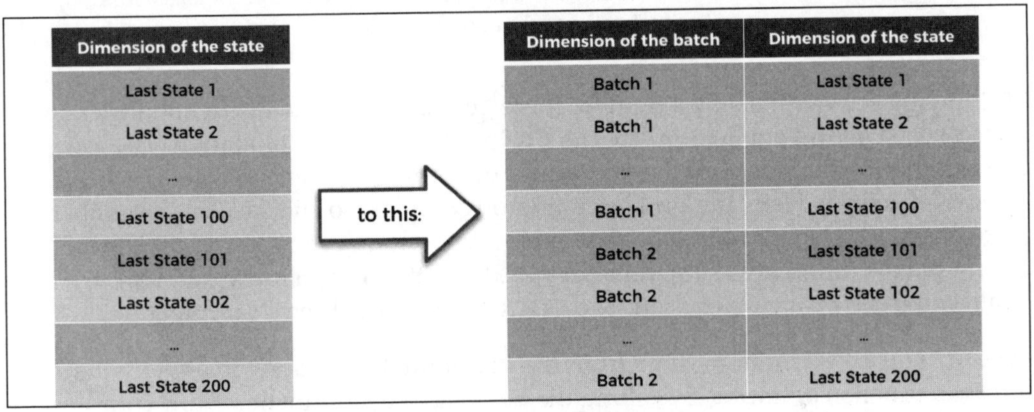

Dimension of the state		Dimension of the batch	Dimension of the state
Last State 1		Batch 1	Last State 1
Last State 2		Batch 1	Last State 2
...	**to this:**	...	...
Last State 100		Batch 1	Last State 100
Last State 101		Batch 2	Last State 101
Last State 102		Batch 2	Last State 102
...		...	...
Last State 200		Batch 2	Last State 200

Figure 12: Adding a dimension for the batch

Line 56: You introduce the sixth object variable, `self.last_action`, initialized as `0`, which is the last action played at each iteration.

Line 57: We introduce the last object variable, `self.last_reward`, initialized as `0`, which is the last reward received after playing the last action `self.last_action`, in the last state `self.last_state`.

Now, you're all good for the __init__ method. Let's move on to the next code section with the next method: the `select_action` method, which selects the action to play at each iteration using Softmax.

```
59          def select_action(self, state):
60              probs = F.softmax(self.model(Variable(state))*100)
61              action = probs.multinomial(len(probs))
```

```
62              return action.data[0,0]
```

Line 59: You start defining the `select_action` method, which takes as input an input state vector (orientation, signal 1, signal 2, signal 3), and returns as output the selected action to play.

Line 60: You get the probabilities of the three actions thanks to the Softmax function taken from the `torch.nn.functional` module. This Softmax function takes the Q-values as input, which are exactly returned by `self.model(Variable(state))`. Remember, `self.model` is an object of the `Network` class, which has the `forward` method, which takes as input an input state tensor wrapped into a `torch Variable`, and returns as output the Q-values for the three actions.

Geek note: Usually we would specify that we call the `forward` method this way – `self.model.forward(Variable(state))` – but since `forward` is the only method of the `Network` class, it is sufficient to just call `self.model`.

Multiplying the Q-values by a number (here `100`) inside `softmax` is a good trick to remember: it allows you to regulate the Exploration versus Exploitation. The lower that number is, the more you'll explore, and therefore the longer it will take to get optimized actions. Here, the problem's not too complex, so choose a large number (`100`) in order to have confident actions and a smooth trajectory to the goal. You'll clearly see the difference if you remove `*100` from the code. Simply put, with the `*100`, you'll see a car sure of itself; without the `*100`, you'll see a car fidgeting.

Line 61: You take a random draw from the distribution of actions created by the `softmax` function at line 60, by calling the `multinomial()` function from your probabilities `probs`.

Line 62: You return the selected action to perform, which you access in `action.data[0,0]`. The returned `action` has an advanced tensor structure, and the action index (0, 1, or 2) that you're interested in is located in the `data` attribute of the action tensor at the first cell of indexes [0,0].

Let's move on to the next code section, the `learn` method. This one is pretty interesting because it's where the heart of deep Q-learning beats. It's in this method that we compute the temporal difference, and accordingly the loss, and update the weights with our optimizer in order to reduce that loss. That's why this method is called `learn`, because it is here that the AI learns to perform better and better actions that increase the accumulated reward. Let's continue:

```
64          def learn(self, batch_states, batch_actions, batch_rewards,
       batch_next_states):
65              batch_outputs = self.model(batch_states).gather(1, batch_
       actions.unsqueeze(1)).squeeze(1)
66              batch_next_outputs = self.model(batch_next_states).
       detach().max(1)[0]
67              batch_targets = batch_rewards + self.gamma * batch_next_
       outputs
68              td_loss = F.smooth_l1_loss(batch_outputs, batch_targets)
69              self.optimizer.zero_grad()
70              td_loss.backward()
71              self.optimizer.step()
```

Line 64: You start by defining the `learn()` method, which takes as inputs the batches of the four elements composing a transition (input state, action, reward, next state):

1. `batch_states`: A batch of input states.
2. `batch_actions`: A batch of actions played.
3. `batch_rewards`: A batch of the rewards received.
4. `batch_next_states`: A batch of the next states reached.

Before I explain Lines 65, 66, and 67, let's take a step back and see what you'll have to do. As you know, the goal of this `learn` method is to update the weights in directions that reduce the back-propagated loss at each iteration of the training. First let's remind ourselves of the formula for the loss:

$$Loss = \frac{1}{2}\sum_{B}\left(R\left(s_{t_B}, a_{t_B}\right) + \gamma \max_{a}\left(Q\left(s_{t_B+1}, a\right)\right) - Q\left(s_{t_B}, a_{t_B}\right) \right)^2 = \frac{1}{2}\sum_{B} TD_{t_B}\left(s_{t_B}, a_{t_B}\right)^2$$

Inside the formula for the loss, we clearly recognize the outputs (predicted Q-values) and the targets:

$$\text{Batch of outputs}: Q\left(s_{t_B}, a_{t_B}\right)$$

$$\text{Batch of targets}: R\left(s_{t_B}, a_{t_B}\right) + \gamma \max_{a}\left(Q\left(s_{t_B+1}, a\right)\right)$$

Therefore, to compute the loss, you proceed this way over the next four lines of code:

Line 65: You collect the batch of outputs, $Q\left(s_{t_B}, a_{t_B}\right)$.

Line 66: You compute the $\max_{a}\left(Q\left(s_{t_{B}+1},a\right)\right)$ part of the targets, which you call `batch_next_outputs`.

Line 67: You get the batch of targets.

Line 68: Since you have the outputs and targets, you're ready to get the loss.

Now let's do this in detail.

Line 65: You collect the batch of outputs $Q\left(s_{t_{B}},a_{t_{B}}\right)$, meaning the predicted Q-values of the input states and the actions played in the batch. Getting them takes several steps. First, you call `self.model(batch_states)`, which, as seen in Line 60, returns the Q-values of each input state in `batch_states` and for all the three actions 0, 1, and 2. To help you visualize it better, it returns something like this:

	Action a0	Action a1	Action a2
Input State s1	Q(s1,a0)	Q(s1,a1)	Q(s1,a2)
Input State s2	Q(s2,a0)	Q(s2,a1)	Q(s2,a2)
...	...	...	...
Input State s100	Q(s100,a0)	Q(s100,a1)	Q(s100,a2)

Figure 13: What is returned by self.model(batch_states)

You only want the predicted Q-values for the selected actions from the batch of outputs, which are found in the batch of actions `batch_actions`. That's exactly what the `.gather(1, batch_actions.unsqueeze(1)).squeeze(1)` trick does: for each input state of the batch, it picks the Q-value that corresponds to the action that was selected in the batch of actions. To help visualize this better, let's suppose the batch of actions is the following:

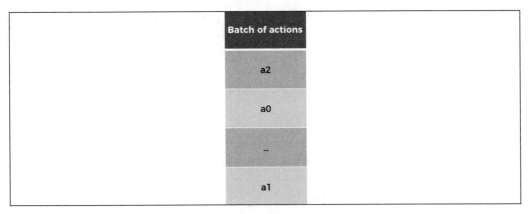

Figure 14: Batch of actions

Then you would get the following batch of outputs composed of the red Q-values:

	Action a0	Action a1	Action a2		Batch of outputs
Input State s1	Q(s1,a0)	Q(s1,a1)	Q(s1,a2)		Q(s1,a2)
Input State s2	Q(s2,a0)	Q(s2,a1)	Q(s2,a2)		Q(s2,a0)
...	...	...	...		...
Input State s100	Q(s100,a0)	Q(s100,a1)	Q(s100,a2)		Q(s100,a1)

Figure 15: Batch of outputs

I hope this is clear; I'm doing my best not to lose you along the way.

Line 66: Now you get the $\max_a\left(Q\left(s_{t_B+1}, a\right)\right)$ part of the target. Call this `batch_next_outputs`; you get it in two steps. First, call `self.model(batch_next_states)` to get the predicted Q-values for each next state of the batch of next states and for each of the three actions. Then, for each next state of the batch, take the maximum of the three Q-values using `.detach().max(1)[0]`. That gives you the batch of the $\max_a\left(Q\left(s_{t_B+1}, a\right)\right)$ values part of the targets.

Line 67: Since you have the batch of rewards $R\left(s_{t_B}, a_{t_B}\right)$ (it's part of the arguments), and since you just got the batch of the $\max\limits_{a}\left(Q\left(s_{t_B+1}, a\right)\right)$ values part of the targets at Line 66, then you're ready to get the batch of targets:

$$\text{Batch of targets}: R\left(s_{t_B}, a_{t_B}\right) + \gamma \max_{a}\left(Q\left(s_{t_B+1}, a\right)\right)$$

That's exactly what you do at Line 67, by summing `batch_rewards` and `batch_next_outputs` multiplied by `self.gamma`, one of the object variables in the Dqn class. Now you have both the batch of outputs and the batch of targets, so you're ready to get the loss.

Line 68: Let's remind ourselves of the formula for the loss:

$$Loss = \frac{1}{2}\sum_{B}\left(R\left(s_{t_B}, a_{t_B}\right) + \gamma \max_{a}\left(Q\left(s_{t_B+1}, a\right)\right) - Q\left(s_{t_B}, a_{t_B}\right)\right)^2$$

$$Loss = \frac{1}{2}\sum_{B}\left(\text{Target} - \text{Output}\right)^2$$

$$Loss = \frac{1}{2}\sum_{B} TD_{t_B}\left(s_{t_B}, a_{t_B}\right)^2$$

Therefore, in order to get the loss, you just have to get the sum of the squared differences between our targets and outputs in the batch. That's exactly what the `smooth_l1_loss` function will do. Taken from the `torch.nn.functional` module, it takes as inputs the two batches of outputs and targets and returns the loss as given in the preceding formula. In the code, call this loss `td_loss` as in **temporal difference loss**.

Excellent progress! Now that you have the loss, representing the error between the predictions and the targets, you're ready to backpropagate this loss into the neural network and update our weights to reduce this loss through mini-batch gradient descent. That's why the next step to take here is to use your optimizer, which is the tool that will perform the updates to the weights.

Line 69: You first initialize the gradients, by calling the `zero_grad()` method from your `self.optimizer` object (`zero_grad` is a method of the Adam class), which will basically set all the gradients of the weights to zero.

Line 70: You backpropagate the loss error `td_loss` into the neural network by calling the `backward()` function from `td_loss`.

Line 71: You perform the weights updates by calling the `step()` method from your `self.optimizer` object (`step` is a method of the `Adam` class).

Congratulations! You've built yourself a tool in the `Dqn` class that will train your car to drive better. You've done the toughest part. Now all you have left to do is to wrap things up into a last method, called `update`, which will simply update the weights after reaching a new state.

Now, in case you are thinking, "but isn't what I've already done with the `learn` method?," well, you're right; but you need to make an extra function that will update the weights at the right time. The right time to update the weights is right after our AI reaches a new state. Simply put, this final `update` method you're about to implement will connect the dots between the `learn` method and the dynamic environment.

That's the finish line! Are you ready? Here's the code:

```
73        def update(self, new_state, new_reward):
74            new_state = torch.Tensor(new_state).float().unsqueeze(0)
75            self.memory.push((self.last_state, torch.
       LongTensor([int(self.last_action)]), torch.Tensor([self.last_
       reward]), new_state))
76            new_action = self.select_action(new_state)
77            if len(self.memory.memory) > 100:
78                batch_states, batch_actions, batch_rewards, batch_
       next_states = self.memory.sample(100)
79                self.learn(batch_states, batch_actions, batch_rewards,
       batch_next_states)
80            self.last_state = new_state
81            self.last_action = new_action
82            self.last_reward = new_reward
83            return new_action
```

Line 73: You introduce the `update()` method, which takes as input the new state reached and the new reward received right after playing an action. This new state entered here will be the `state` variable you can see in Line 129 of the `map.py` file and this new reward will be the `reward` variable you can see in Lines 138 to 145 of the `map.py` file. This `update` method performs some operations including the weights updates and, in the end, returns the new action to perform.

Line 74: You first convert the new state into a torch tensor and unsqueeze it to create an additional dimension (placed first in index 0) corresponding to the batch. To ease future operations, you also make sure that all the elements of the new state (orientation plus the three signals) are converted into floats by adding `.float()`.

Line 75: Using the `push()` method from your memory object, add a new transition to the memory. This new transition is composed of:

1. `self.last_state`: The last state reached before reaching that new state
2. `self.last_action`: The last action played that led to that new state
3. `self.last_reward`: The last reward received after performing that last action
4. `new_state`: The new state that was just reached

All the elements of this new transition are converted into torch tensors.

Line 76: Using the `select_action()` method from your `Dqn` class, perform a new action from the new state just reached.

Line 77: Check if the size of the memory is larger than 100. In `self.memory.memory`, the first `memory` is the object created at Line 53 and the second `memory` is the variable object introduced at Line 35.

Line 78: If that's the case, sample 100 transitions from the memory, using the `sample()` method from your `self.memory` object. This returns four batches of size 100:

1. `batch_states`: The batch of current states (current at the time of the transition).
2. `batch_actions`: The batch of actions performed in the current states.
3. `batch_rewards`: The batch of rewards received after playing the actions of `batch_actions` in the current states of `batch_states`.
4. `batch_next_states`: The batch of next states reached after playing the actions of `batch_actions` in the current states of `batch_states`.

Line 79: Still in the `if` condition, proceed to the weights updates using the `learn()` method called from the same `Dqn` class, with the four previous batches as inputs.

Line 80: Update the last state reached, `self.last_state`, which becomes `new_state`.

Line 81: Update the last action performed, `self.last_action`, which becomes `new_action`.

Line 82: Update the last reward received, `self.last_reward`, which becomes `new_reward`.

Line 83: Return the new action performed.

That's it for the `update()` method! I hope you can see how we connected the dots. Now, to connect the dots even better, let's see where and how you call that `update` method in the `map.py` file.

First, before calling that `update()` method, you have to create an object of the `Dqn` class, which here is called `brain`. That's exactly what you do in Line 33 of the `map.py` file.

```
33      brain = Dqn(4,3,0.9)
```

The arguments entered here are the three arguments we see in the `__init__()` method of the `Dqn` class:

- `4` is the number of elements in the input state (`input_size`).
- `3` is the number of possible actions (`nb_action`).
- `0.9` is the discount factor (`gamma`).

Then, from this `brain` object, you call on the `update()` method in Line 130 of the `map.py` file, right after reaching a new state, called `state` in the code:

```
129           state = [orientation, self.car.signal1, self.car.
        signal2, self.car.signal3]
130           action = brain.update(state, reward)
```

Going back to your `Dqn` class, you need two extra methods:

1. The `save()` method, which saves the weights of the AI's network after their last updates. This method will be called as soon as you click the **save** button while running the map. The weights of your AI will be then saved and put into a file named `last_brain.pth`, which will automatically be populated in the folder that contains your Python files. That's what allows you to have a pre-trained AI.

2. The `load()` method, which loads the saved weights in the `last_brain.pth` file. This method will be called as soon as you click the **load** button while running the map. It allows you to start the map with a pre-trained self-driving car, without having to wait for it to train.

These last two methods aren't AI-related, so we won't spend time explaining each line of their code. Still, it's good for you to be able to recognize these two tools in case you want to use them for another AI model that you build with PyTorch.

Here's how they're implemented:

```
85      def save(self):
86          torch.save({'state_dict': self.model.state_dict(),
87                      'optimizer' : self.optimizer.state_dict(),
88                      }, 'last_brain.pth')
89
90      def load(self):
91          if os.path.isfile('last_brain.pth'):
92              print("=> loading checkpoint... ")
93              checkpoint = torch.load('last_brain.pth')
94              self.model.load_state_dict(checkpoint['state_dict'])
95              self.optimizer.load_state_
    dict(checkpoint['optimizer'])
96              print("done !")
97          else:
98              print("no checkpoint found...")
```

Congratulations!

That's right! You've finished this 100 lines of code implementation of the AI inside our self-driving car. That's quite an accomplishment, especially when coding deep Q-learning for the first time. You really can be proud to have gone this far.

After all this hard work, you definitely deserve to have some fun, and I think it'll be the most fun to watch the result of your hard work. In other words, you're about to see your self-driving car in action! I remember I was so excited the first time I ran this. You'll feel it too; it's pretty cool!

The demo

I have some good news and some bad news.

I'll start with the bad news: we can't run the map.py file with a simple plug and play on Google Colab. The reason for that is that Kivy is very tricky to install through Colab. So, we'll go for the classic method of running a Python file: through the terminal.

The good news is that once we install Kivy and PyTorch through the terminal, you'll have a fantastic demo!

Let's install everything we need to run our self-driving car. Here's what we have to install, in the following order:

1. **Anaconda**: A free and open source distribution of Python that offers an easy way to install packages thanks to the `conda` command. This is what we'll use to install PyTorch and Kivy.

2. **Virtual environment with Python 3.6**: Anaconda is installed with Python 3.7 or higher; however, that 3.7 version is not compatible with Kivy. We'll create a virtual environment in which we install Python 3.6, a version compatible with both Kivy and our implementation. Don't worry if that sounds intimidating, I'll give you all the details you need to set this up.

3. **PyTorch**: Then, inside the virtual environment, we'll install PyTorch, the AI framework used to build our deep Q-network. We'll install a specific version of PyTorch that's compatible with our implementation, so that everyone can be on the same page and run it with no issues. PyTorch upgrades sometimes include changes in the names of the modules, which can make an old implementation incompatible with the newest PyTorch versions. Here, we know we have the right PyTorch version for our implementation.

4. **Kivy**: To finish, still inside the virtual environment, we'll install Kivy, the open source Python framework on which we will run our map.

Let's start with Anaconda.

Installing Anaconda

On Google, or your favorite browser, go to www.anaconda.com. On the Anaconda website, click **Download** on the upper right corner of the screen. Scroll down and you'll find the Python versions to download:

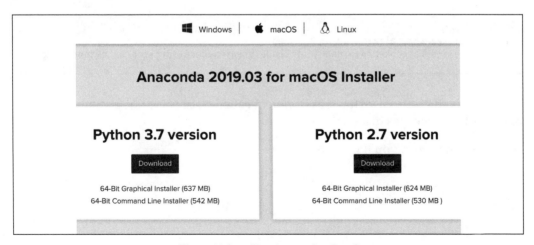

Figure 16: Installing Anaconda – Step 2

At the top, make sure that your system (Windows, macOS, or Linux) is correctly selected. If it is, click the **Download** button in the Python 3.7 version box. This will download Anaconda with Python 3.7.

Then double-click the downloaded file and keep clicking **Continue** and **Agree** to install, until the end. If you're prompted to choose who or how to install it for, choose **install for me only**.

Creating a virtual environment with Python 3.6

Now that Anaconda's installed, you can create a virtual environment, named `selfdrivingcar`, with Python 3.6 installed. To do this you need to open a terminal and enter some commands. Here's how to open it for the three systems:

1. For Linux users, just press *Ctrl + Alt + T*.

2. For Mac users, press *Cmd + Space*, and then in the Spotlight Search enter `Terminal`.

3. For Windows users, click the Windows button at the lower left corner of your screen, find `anaconda` in the list of programs, and click to open Anaconda prompt. A black window will open; that's the terminal you'll use to install the packages.

Inside the terminal, enter the following command:

```
conda create -n selfdrivingcar python=3.6
```

Just like so:

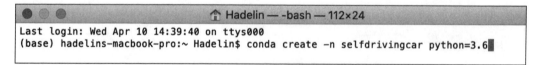

This command creates a virtual environment called `selfdrivingcar` with Python 3.6 and other packages installed.

After pressing *Enter*, you'll get this in a few seconds:

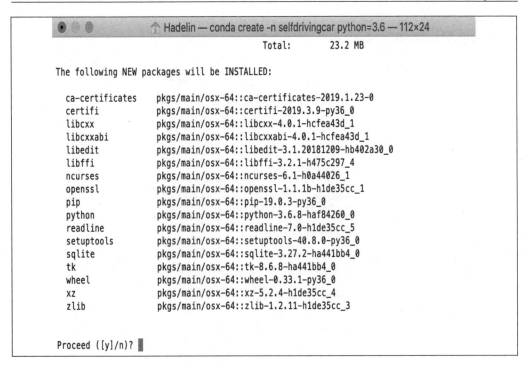

Press y to proceed. This will download and extract the packages. After a few seconds, you'll get this, which marks the end of the installation:

```
●  ○  ○                      ⌂ Hadelin — -bash — 112×24

    Proceed ([y]/n)? y


    Downloading and Extracting Packages
    python-3.6.8         | 20.5 MB   | ################################################################ | 100%
    wheel-0.33.1         | 39 KB     | ################################################################ | 100%
    pip-19.0.3           | 1.9 MB    | ################################################################ | 100%
    certifi-2019.3.9     | 155 KB    | ################################################################ | 100%
    setuptools-40.8.0    | 622 KB    | ################################################################ | 100%
    Preparing transaction: done
    Verifying transaction: done
    Executing transaction: done
    #
    # To activate this environment, use
    #
    #     $ conda activate selfdrivingcar
    #
    # To deactivate an active environment, use
    #
    #     $ conda deactivate

    (base) hadelins-macbook-pro:~ Hadelin$ ▌
```

Then we're going to activate the selfdrivingcar virtual environment, meaning we're going to get inside it in order to install PyTorch and Kivy within the selfdrivingcar virtual environment.

As you can see just preceding, to activate the environment, we will enter the following command:

```
conda activate selfdrivingcar
```

Enter that command, and then you'll get inside the virtual environment:

```
● ● ●                              ⌂ Hadelin — -bash — 112×24

Proceed ([y]/n)? y


Downloading and Extracting Packages
python-3.6.8       | 20.5 MB  | ################################################################## | 100%
wheel-0.33.1       | 39 KB    | ################################################################## | 100%
pip-19.0.3         | 1.9 MB   | ################################################################## | 100%
certifi-2019.3.9   | 155 KB   | ################################################################## | 100%
setuptools-40.8.0  | 622 KB   | ################################################################## | 100%
Preparing transaction: done
Verifying transaction: done
Executing transaction: done
#
# To activate this environment, use
#
#     $ conda activate selfdrivingcar
#
# To deactivate an active environment, use
#
#     $ conda deactivate

[(base) hadelins-macbook-pro:~ Hadelin$ conda activate selfdrivingcar                                       ]
(selfdrivingcar) hadelins-macbook-pro:~ Hadelin$ █
```

Now we can see (selfdrivingcar) before my computer's name, hadelins-macbook-pro, which means we are inside the selfdrivingcar virtual environment.

We're ready for the next steps, which are the installation of PyTorch and Kivy inside this virtual environment. Don't close your terminal, or when you open it again you'll be back in the main environment.

Installing PyTorch

Now we're going to install PyTorch inside the virtual environment by entering the following command:

```
conda install pytorch==0.3.1 -c pytorch
```

Just like so:

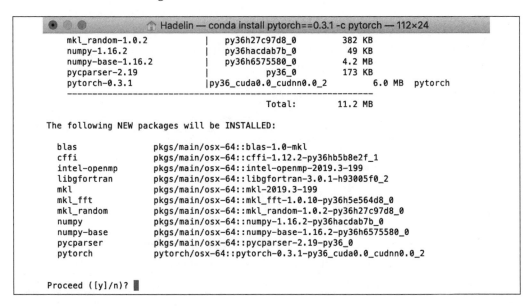

```
                      🔵 Hadelin — -bash — 112×24
  
  Proceed ([y]/n)? y
  
  
  Downloading and Extracting Packages
  python-3.6.8       | 20.5 MB   | ############################################################ | 100%
  wheel-0.33.1       | 39 KB     | ############################################################ | 100%
  pip-19.0.3         | 1.9 MB    | ############################################################ | 100%
  certifi-2019.3.9   | 155 KB    | ############################################################ | 100%
  setuptools-40.8.0  | 622 KB    | ############################################################ | 100%
  Preparing transaction: done
  Verifying transaction: done
  Executing transaction: done
  #
  # To activate this environment, use
  #
  #     $ conda activate selfdrivingcar
  #
  # To deactivate an active environment, use
  #
  #     $ conda deactivate
  
  [(base) hadelins-macbook-pro:~ Hadelin$ conda activate selfdrivingcar                               ]
  .(selfdrivingcar) hadelins-macbook-pro:~ Hadelin$ conda install pytorch==0.3.1 -c pytorch█
```

After a few seconds, we get this:

```
                  🔵 Hadelin — conda install pytorch==0.3.1 -c pytorch — 112×24
     mkl_random-1.0.2      |     py36h27c97d8_0         382 KB
     numpy-1.16.2          |     py36hacdab7b_0          49 KB
     numpy-base-1.16.2     |     py36h6575580_0         4.2 MB
     pycparser-2.19        |          py36_0            173 KB
     pytorch-0.3.1         |py36_cuda0.0_cudnn0.0_2      6.0 MB   pytorch
     -------------------------------------------------------------
                                       Total:          11.2 MB

  The following NEW packages will be INSTALLED:

     blas              pkgs/main/osx-64::blas-1.0-mkl
     cffi              pkgs/main/osx-64::cffi-1.12.2-py36hb5b8e2f_1
     intel-openmp      pkgs/main/osx-64::intel-openmp-2019.3-199
     libgfortran       pkgs/main/osx-64::libgfortran-3.0.1-h93005f0_2
     mkl               pkgs/main/osx-64::mkl-2019.3-199
     mkl_fft           pkgs/main/osx-64::mkl_fft-1.0.10-py36h5e564d8_0
     mkl_random        pkgs/main/osx-64::mkl_random-1.0.2-py36h27c97d8_0
     numpy             pkgs/main/osx-64::numpy-1.16.2-py36hacdab7b_0
     numpy-base        pkgs/main/osx-64::numpy-base-1.16.2-py36h6575580_0
     pycparser         pkgs/main/osx-64::pycparser-2.19-py36_0
     pytorch           pytorch/osx-64::pytorch-0.3.1-py36_cuda0.0_cudnn0.0_2


  Proceed ([y]/n)? █
```

Press y again, and then press *Enter*.

After a few seconds, PyTorch is installed:

```
●  ●  ●                          🏠 Hadelin — -bash — 112×24
    numpy                pkgs/main/osx-64::numpy-1.17.2-py36h99e6662_0
    numpy-base           pkgs/main/osx-64::numpy-base-1.17.2-py36h6575580_0
    pycparser            pkgs/main/osx-64::pycparser-2.19-py36_0
    pytorch              pytorch/osx-64::pytorch-0.3.1-py36_cuda0.0_cudnn0.0_2
    six                  pkgs/main/osx-64::six-1.12.0-py36_0


Proceed ([y]/n)? y


Downloading and Extracting Packages
numpy-base-1.17.2    | 4.0 MB   | ##################################################################### | 100%
mkl_random-1.1.0     | 287 KB   | ##################################################################### | 100%
pycparser-2.19       | 170 KB   | ##################################################################### | 100%
cffi-1.12.3          | 212 KB   | ##################################################################### | 100%
mkl-service-2.3.0    | 202 KB   | ##################################################################### | 100%
numpy-1.17.2         | 4 KB     | ##################################################################### | 100%
pytorch-0.3.1        | 6.0 MB   | ##################################################################### | 100%
mkl_fft-1.0.14       | 139 KB   | ##################################################################### | 100%
six-1.12.0           | 22 KB    | ##################################################################### | 100%
Preparing transaction: done
Verifying transaction: done
Executing transaction: done
(selfdrivingcar) hadelins-macbook-pro:~ Hadelin$ ▎
```

Installing Kivy

Now let's proceed to Kivy. In the same virtual environment, we're going to install Kivy by entering the following command:

```
conda install -c conda-forge/label/cf201901 kivy
```

```
●  ●  ●                          🏠 Hadelin — -bash — 112×24
   numpy                pkgs/main/osx-64::numpy-1.17.2-py36h99e6662_0
   numpy-base           pkgs/main/osx-64::numpy-base-1.17.2-py36h6575580_0
   pycparser            pkgs/main/osx-64::pycparser-2.19-py36_0
   pytorch              pytorch/osx-64::pytorch-0.3.1-py36_cuda0.0_cudnn0.0_2
   six                  pkgs/main/osx-64::six-1.12.0-py36_0


Proceed ([y]/n)? y


Downloading and Extracting Packages
numpy-base-1.17.2    | 4.0 MB   | ##################################################################### | 100%
mkl_random-1.1.0     | 287 KB   | ##################################################################### | 100%
pycparser-2.19       | 170 KB   | ##################################################################### | 100%
cffi-1.12.3          | 212 KB   | ##################################################################### | 100%
mkl-service-2.3.0    | 202 KB   | ##################################################################### | 100%
numpy-1.17.2         | 4 KB     | ##################################################################### | 100%
pytorch-0.3.1        | 6.0 MB   | ##################################################################### | 100%
mkl_fft-1.0.14       | 139 KB   | ##################################################################### | 100%
six-1.12.0           | 22 KB    | ##################################################################### | 100%
Preparing transaction: done
Verifying transaction: done
Executing transaction: done
(selfdrivingcar) hadelins-macbook-pro:~ Hadelin$ conda install -c conda-forge/label/cf201901 kivy▎
```

Again, we get this:

```
● ● ●                    🏠 Hadelin — conda install -c conda-forge kivy — 112×24
  libiconv              conda-forge/osx-64::libiconv-1.15-h01d97ff_1005
  libpng               conda-forge/osx-64::libpng-1.6.36-ha441bb4_1000
  libtiff              conda-forge/osx-64::libtiff-4.0.10-h79f4b77_1001
  olefile              conda-forge/noarch::olefile-0.46-py_0
  pcre                 conda-forge/osx-64::pcre-8.41-h0a44026_1003
  pillow               conda-forge/osx-64::pillow-5.2.0-py36h2dc6135_1
  pygments             conda-forge/noarch::pygments-2.3.1-py_0
  sdl2                 conda-forge/osx-64::sdl2-2.0.8-h0a44026_1001
  sdl2_image           conda-forge/osx-64::sdl2_image-2.0.4-hacdbef4_0
  sdl2_mixer           conda-forge/osx-64::sdl2_mixer-2.0.1-hfc679d8_1
  sdl2_ttf             conda-forge/osx-64::sdl2_ttf-2.0.14-hdd9f355_1
  smpeg2               conda-forge/osx-64::smpeg2-2.0.0-hfc679d8_1

The following packages will be UPDATED:

  ca-certificates      pkgs/main::ca-certificates-2019.1.23-0 --> conda-forge::ca-certificates-2019.3.9-hecc5488_0
  openssl              pkgs/main::openssl-1.1.1b-h1de35cc_1 --> conda-forge::openssl-1.1.1b-h01d97ff_2

The following packages will be SUPERSEDED by a higher-priority channel:

  certifi                                    pkgs/main --> conda-forge

Proceed ([y]/n)? █
```

Enter y again, and after a few seconds more, Kivy is installed.

```
● ● ●                    🏠 Hadelin — -bash — 112×24
kivy-1.10.1          | 21.3 MB | ######################################################### | 100%
ca-certificates-2019 | 146 KB  | ######################################################### | 100%
pcre-8.41            | 222 KB  | ######################################################### | 100%
sdl2_mixer-2.0.1     | 151 KB  | ######################################################### | 100%
olefile-0.46         | 31 KB   | ######################################################### | 100%
libtiff-4.0.10       | 486 KB  | ######################################################### | 100%
openssl-1.1.1b       | 3.5 MB  | ######################################################### | 100%
freetype-2.8.1       | 830 KB  | ######################################################### | 100%
certifi-2019.3.9     | 149 KB  | ######################################################### | 100%
smpeg2-2.0.0         | 158 KB  | ######################################################### | 100%
gettext-0.19.8.1     | 3.3 MB  | ######################################################### | 100%
libiconv-1.15        | 1.3 MB  | ######################################################### | 100%
glib-2.58.3          | 3.1 MB  | ######################################################### | 100%
jpeg-9c              | 237 KB  | ######################################################### | 100%
pygments-2.3.1       | 641 KB  | ######################################################### | 100%
gst-plugins-base-1.1 | 2.1 MB  | ######################################################### | 100%
gstreamer-1.14.4     | 1.7 MB  | ######################################################### | 100%
libpng-1.6.36        | 306 KB  | ######################################################### | 100%
pillow-5.2.0         | 542 KB  | ######################################################### | 100%
sdl2_ttf-2.0.14      | 29 KB   | ######################################################### | 100%
Preparing transaction: done
Verifying transaction: done
Executing transaction: done
(selfdrivingcar) hadelins-macbook-pro:~ Hadelin$ █
```

Now I have some terrific news for you: you're ready to run the self-driving car! To do that, we need to run our code in the terminal, still inside our virtual environment.

 If you already closed your terminal, then when you open it again enter the conda activate selfdrivingcar command in order to get back inside the virtual environment.

So, let's run the code! If you haven't already, download the whole repository by clicking the **Clone or download** button on the GitHub page:

(`https://github.com/PacktPublishing/AI-Crash-Course`)

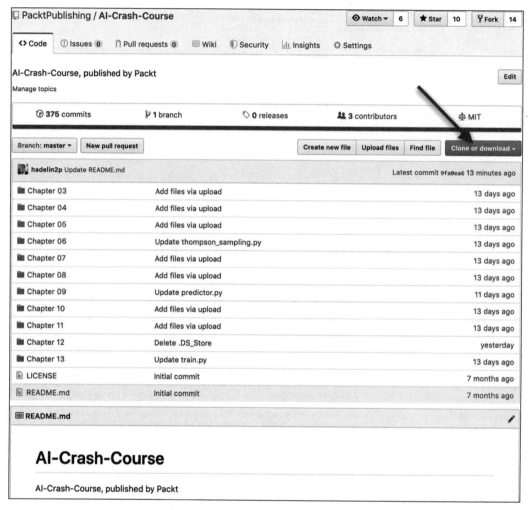

Figure 17: The GitHub repository

Then unzip it and move the unzipped folder to your desktop, just like so:

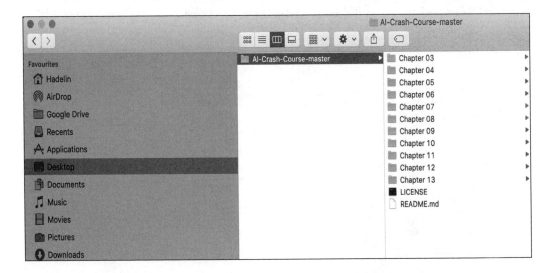

Now go into `Chapter 10` and select and copy all the files inside:

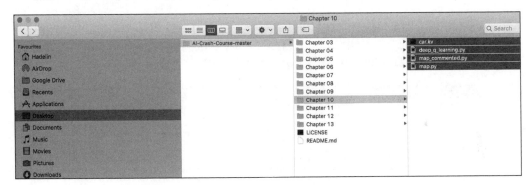

Then, because we're only interested in these files right now, and to simplify the command lines in the terminal, paste these files inside the main `AI-Crash-Course-master` folder and remove all the rest, which we don't need, so that you eventually end up with this:

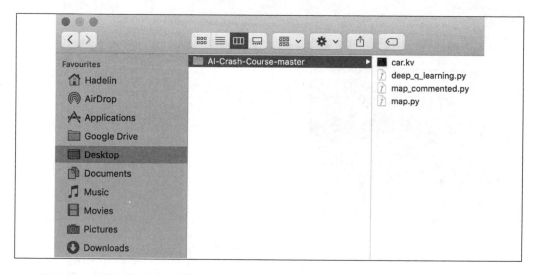

Now we're going to access this folder from the terminal. Since we put the repository folder in the desktop, we will find it in a flash. Back into the terminal, enter `ls` (l as in lion) to see in which folder you are in your machine:

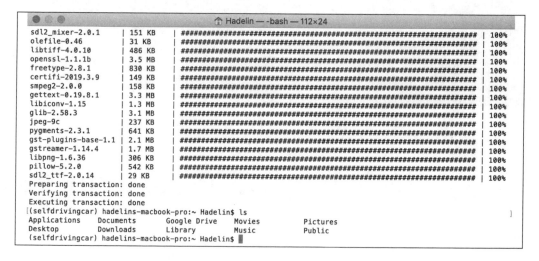

I can see that I'm in my main root folder, which contains the `Desktop` folder. It should usually be the case for you too. So now we're going to go into the `Desktop` folder by entering the following command:

```
cd Desktop
```

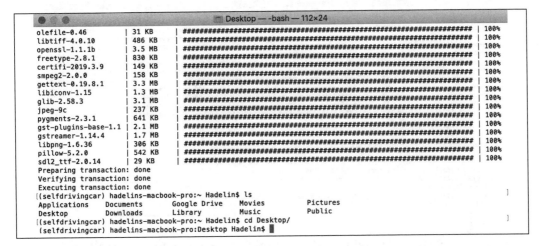

Enter `ls` again and check that you indeed see the `AI-Crash-Course-master` folder:

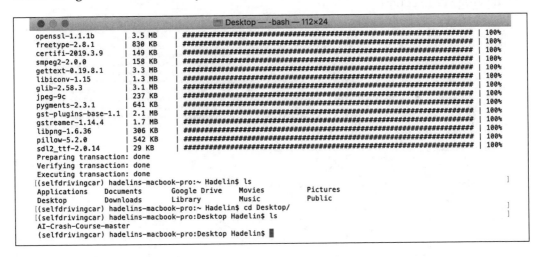

Then go into the `AI-Crash-Course-master` folder by entering the following command:

```
cd AI-Crash-Course-master
```

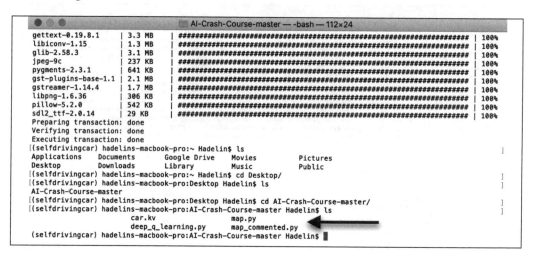

Perfect! Now we're in the right spot! By entering `ls` again, you can see all the files of the repo, including the `map.py` file, which is the one we have to run to see our self-driving car in action!

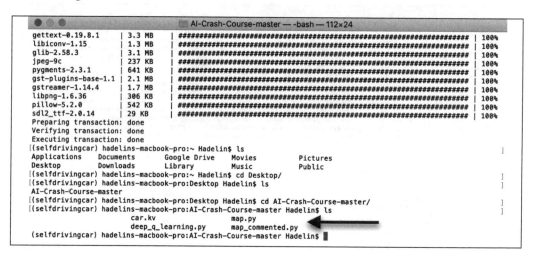

If by any chance you had trouble getting to this point, that may be because your main root folder doesn't contain your `Desktop` folder. If that's the case, just put the `AI-Crash-Course-master` repo folder inside one of the folders that you see when entering the `ls` command in the terminal, and redo the same process.

What you have to do is just find and enter the `AI-Crash-Course-master` folder with the `cd` commands. That's it! Don't forget to make sure your `AI-Crash-Course-master` folder only contains the self-driving car files:

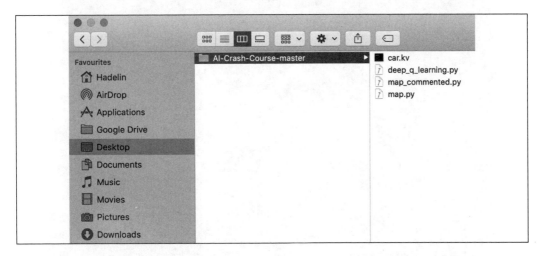

Now you're only one command line away from running your self-driving car. I hope you're excited to see the results of your hard work; I know exactly how you feel, I was in your shoes not so long ago!

So, without further ado, let's enter the final command, right now. It's this:

```
python map.py
```

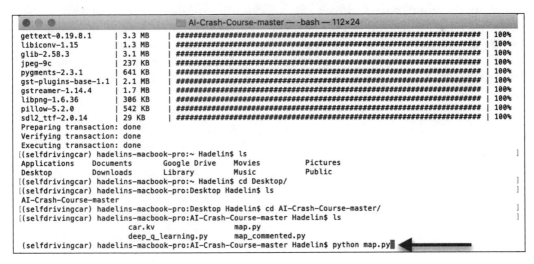

As soon as you enter it, the map with the car will pop up just like so:

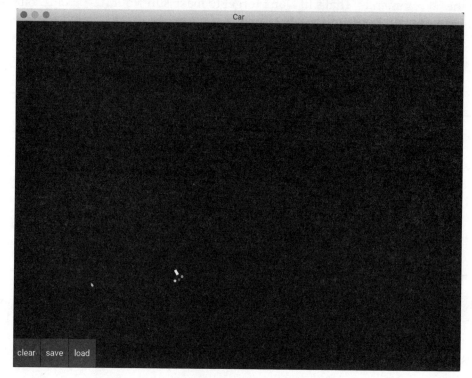

Figure 18: The map

For the first minute or so, your self-driving car will explore its actions by performing nonsense movements; you might see it spinning around. After each 100 movements, the weights inside the neural network of the AI get updated, and the car improves its actions to get higher rewards. And suddenly, maybe after another 30 seconds or so, you should see your car making round trips between the Airport and Downtown, which I highlighted here again:

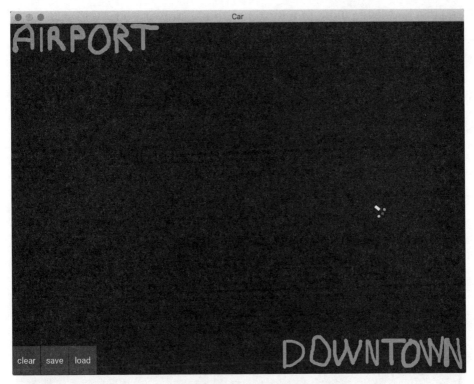

Figure 19: The destinations

Now have some fun! Draw some obstacles on the map to see if the car avoids them.

On my side I have just drawn this, and after a few more minutes of training, I can clearly see the car avoiding the obstacles:

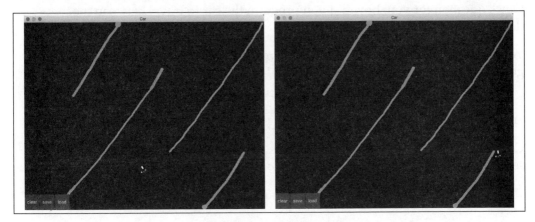

Figure 20: Road with obstacles

And you can have even more fun! By, for example, drawing a road like so:

Figure 21: The road of the demo

After a few minutes of training, the car becomes able to self-drive along that road, while making many road trips between the Airport and Downtown.

Quick question for you: how did you program the car to travel between the destinations?

You did it by giving a small positive reward to the AI when the car gets closer to the goal. That's programmed in rows 144 and 145 inside the map.py file:

```
144                 if distance < last_distance:
145                     reward = 0.1
```

Congratulations to you for completing this massive chapter on this not-so-basic self-driving car application! I hope you had fun, and that you feel proud to have mastered such an advanced model in deep reinforcement learning.

Summary

In this chapter, we learned how to build a deep Q-learning model to drive a self-driving car. As inputs it took the information from the three sensors and its current orientation. As outputs it decided the Q-values for each of the actions of going straight, turning left, or turning right. As for the rewards, we punished it badly for hitting the sand, punished it slightly for going in the wrong direction, and rewarded it slightly for going in the right direction. We made the AI implementation in PyTorch and used Kivy for the graphics. To run all of this we used the Anaconda environment.

Now take a long break, you deserve it! I'll see you in the next chapter for our next AI challenge, where this time we will solve a real-world business problem with cost implications running into the millions.

11
AI for Business – Minimize Costs with Deep Q-Learning

It's great that you can implement a deep Q-learning model to build a self-driving car. Really, once again, huge congratulations to you for that. But I also want you to be able to use deep Q-learning to solve a real-world business problem. With this next application, you'll be more than ready to add value to your work or business by leveraging AI. Even though we'll once again use a specific application, this chapter will provide you with a general AI framework, a blueprint containing the general steps of the process you have to follow when solving a real-world problem with deep Q-learning. This chapter is very important to you and for your career; I don't want you to close this book before you feel confident with the skills you'll learn here. Let's smash this next application together!

Problem to solve

When I said we were going to solve a real-world business problem, I didn't overstate the problem; the problem we're about to tackle with deep Q-learning is very similar to the following, which was solved in the real world via deep Q-learning.

In 2016, DeepMind AI minimized a big part of Google's yearly costs by reducing the Google Data Center's cooling bill by 40% using their DQN AI model (deep Q-learning). Check the link here:

https://deepmind.com/blog/deepmind-ai-reduces-google-data-centre-cooling-bill-40/

In this case study, we'll do something very similar. We'll set up our own server environment, and we'll build an AI that controls the cooling and heating of the server so that it stays in an optimal range of temperatures while using the minimum of energy, therefore minimizing the costs.

Just as the DeepMind AI did, our goal will be to achieve at least 40% energy savings! Are you ready for this? Let's bring it on!

As ever, my first question to you is: What's our first step?

I'm sure by this point I don't need to spell out the answer. Let's get straight to building our environment!

Building the environment

Before we define the states, actions, and rewards, we need to set up the server and explain how it operates. We'll do that in several steps:

1. First, we'll list all the environment parameters and variables by which the server is controlled.

2. After that we'll set the essential assumptions of the problem, on which your AI will rely to provide a solution.

3. Then we'll specify how you'll simulate the whole process.

4. Finally, we'll explain the overall functioning of the server, and how the AI plays its role.

Parameters and variables of the server environment

Here is a list of all the parameters, which keep their values fixed, of the server environment:

1. The average atmospheric temperature for each month.

2. The optimal temperature range of the server, which we'll set as $[18°C, 24°C]$.

3. The minimum temperature, below which the server fails to operate, which we'll set as $-20°C$.

4. The maximum temperature, above which the server fails to operate, which we'll set as $80°C$.

5. The minimum number of users in the server, which we'll set as 10.

6. The maximum number of users in the server, which we'll set as 100.

7. The maximum change of users in the server per minute, which we'll set as 5; so every minute, the server can only have a change of 5 extra users or 5 fewer users at most.

8. The minimum rate of data transmission in the server, which we'll set as 20.

9. The maximum rate of data transmission in the server, which we'll set as 300.

10. The maximum change of the rate of data transmission per minute, which we'll set as 10; so every minute, the rate of data transmission can only change by a maximum value of 10 in either direction.

Next, we'll list all the variables, which have values that fluctuate over time, of the server environment:

1. The temperature of the server at a given minute.

2. The number of users connected to the server at a given minute.

3. The rate of data transmission at a given minute.

4. The energy spent by the AI onto the server (to cool it down or heat it up) at a given minute.

5. The energy that would be spent by the server's integrated cooling system to automatically bring the server's temperature back to the optimal range, whenever the server's temperature goes outside this optimal range. This is to keep track of how much energy a **non-AI** system would use, so we can compare our AI system to it.

All these parameters and variables will be part of the environment, and will influence the actions of our AI.

Next, we'll explain the two core assumptions of the environment. It's important to understand that these assumptions are not AI related, but just used to simplify the environment so that we can focus on creating a functional AI solution.

Assumptions of the server environment

We'll rely on the following two essential assumptions:

Assumption 1 – We can approximate the server temperature

The temperature of the server can be approximated through Multiple Linear Regression, that is, by a linear function of the atmospheric temperature, the number of users and the rate of data transmission, like so:

server temperature = b_0 + b_1 × *atmospheric temperature* + b_2 × *number of users* + b_3 × *rate of data transmission*

where $b_0 \in \mathbb{R}$, $b_1 > 0$, $b_2 > 0$, and $b_3 > 0$.

The raison d'être of this assumption and the reason why $b_1 > 0$, $b_2 > 0$, and $b_3 > 0$ are intuitive to understand. It makes sense that when the atmospheric temperature increases, the temperature of the server increases. The more users that are connected to the server, the more energy the server has to spend handling them, and therefore the higher the temperature of the server will be. Finally, the more data is transmitted inside the server, the more energy the server has to spend processing it, and therefore the higher the temperature of the server will be.

For simplicity's sake, we can just suppose that these correlations are linear. However, you could absolutely run the same simulation by assuming they were quadratic or logarithmic, and altering the code to reflect those equations. This is just my simulation of a virtual server environment; feel free to tweak it as you like!

Let's assume further that after performing this Multiple Linear Regression, we obtained the following values of the coefficients: $b_0 = 0$, $b_1 = 1$, $b_2 = 1.25$, and $b_3 = 1.25$. Accordingly:

server temperature = atmospheric temperature $+ 1.25 \times$ *number of users* $+ 1.25 \times$ *rate of data transmission*

Now, if we were facing this problem in real life, we could get the dataset of temperatures for our server and calculate these values directly. Here, we're just assuming values that are easy to code and understand, because our goal in this chapter is not to perfectly model a real server; it's to go through the steps of solving a real-world problem with AI.

Assumption 2 – We can approximate the energy costs

The energy spent by any cooling system, either our AI or the server's integrated cooling system that we'll compare our AI to, that changes the server's temperature from T_t to T_{t+1} within 1 unit of time (in our case 1 minute), can be approximated again through regression by a linear function of the server's absolute temperature change, as so:

$$E_t = \alpha |\Delta T_t| + \beta = \alpha |T_{t+1} - T_t| + \beta$$

where:

1. E_t is the energy spent by the system on the server between times t and $t+1$ minute.

2. ΔT_t is the change in the server's temperature caused by the system, between times t and $t+1$ minute.

3. T_t is the temperature of the server at time t.

4. T_{t+1} is the temperature of the server at time $t+1$ minute.

5. $\alpha > 0$.

6. $\beta \in \mathbb{R}$.

Let's explain why it intuitively makes sense to make this assumption with $\alpha > 0$. That's simply because the more the AI or the old-fashioned integrated cooling system heats up or cools down the server, the more energy it spends to achieve that heat transfer.

For example, imagine the server suddenly has overheating issues and just reached 80°C; then within one unit of time (1 minute), either system will need much more energy to bring the server's temperature back to its optimal temperature, 24°C, than to bring it back to 50°C.

For simplicity's sake, in this example we suppose that these correlations are linear, instead of calculating true values from a real dataset. In case you're wondering why we take the absolute value, that's simply because when the AI cools down the server, $T_{t+1} < T_t$, so $\Delta T < 0$. Since an energy cost is always positive, we have to take the absolute value of ΔT.

Keeping our desired simplicity in mind, we'll assume that the results of the regression are $\alpha = 1$ and $\beta = 0$, so that we get the following final equation based on Assumption 2:

$$E_t = |\Delta T_t| = |T_{t+1} - T_t|$$

thus:

$E_t = T_{t+1} - T_t$ if $T_{t+1} > T_t$, that is, if the server is heated up,

$E_t = T_t - T_{t+1}$ if $T_{t+1} < T_t$, that is, if the server is cooled down.

Now we've got our assumptions covered, let's explain how we'll simulate the operation of the server, with users logging on and off and data coming in and out.

Simulation

The number of users and the rate of data transmission will randomly fluctuate, to simulate the unpredictable user activity and data requirements of an actual server. This leads to randomness in the temperature. The AI needs to learn how much cooling or heating power it should transfer to the server so as to not deteriorate the server performance, and at the same time, expend as little energy as possible by optimizing its heat transfer.

Now that we have the full picture, I'll explain the overall functioning of the server and the AI inside this environment.

Overall functioning

Inside a data center, we're dealing with a specific server that is controlled by the parameters and variables listed previously. Every minute, some new users log on to the server and some current users log off, therefore updating the number of active users in the server. Also, every minute some new data is transmitted into the server, and some existing data is transmitted outside the server, therefore updating the rate of data transmission happening inside the server.

Hence, based on *Assumption 1* given earlier, the temperature of the server is updated every minute. Now please focus, because this is where you'll understand the huge role the AI has to play on the server.

Two possible systems can regulate the temperature of the server: the AI, or the server's integrated cooling system. The server's integrated cooling system is an unintelligent system that automatically brings the server's temperature back inside its optimal temperature range.

Every minute, the server's temperature is updated. If the server is using the integrated cooling system, that system watches to see what happens; that update can either leave the temperature within the range of optimal temperatures $\left([18°C, 24°C] \right)$, or move it outside this range. If it goes outside the optimal range, for example to 30° C, the server's integrated cooling system automatically brings the temperature back to the closest bound of the optimal range, in this case 24°C. For the purposes of our simulation, we're assuming that no matter how big the change in temperature is, the integrated cooling system can bring it back into the optimal range in under a minute. This is, obviously, an unrealistic assumption, but the purpose of this chapter is for you to build a functioning AI capable of solving the problem, not to perfectly simulate the thermal dynamics of a real server. Once we've completed our example together, I highly recommend that you tinker with the code and try to make it more realistic; for now, to keep things simple, we'll believe in our magically effective integrated cooling system.

If the server is instead using the AI, then in that case the server's integrated cooling system is deactivated and it is the AI itself that updates the temperature of the server to regulate it the best way. The AI changes the temperature after making some prior predictions, not in a purely deterministic way as with the unintelligent integrated cooling system. Before there's an update to the number of users and the rate of data transmission, causing a change in the temperature of the server, the AI predicts if it should cool down the server, do nothing, or heat up the server, and acts. Then the temperature change happens and the AI reiterates.

Since these two systems are distinct from one another, we can evaluate them separately to compare their performance; to train or run the AI on a server, while keeping track of how much energy the integrated cooling system would have used in the same circumstances.

That brings us to the energy. Remember that one primary goal of the AI is to lower the energy cost of running this server. Accordingly, our AI has to try and use less energy than the unintelligent cooling system would use on the server. Since, based on *Assumption 2* given preceding, the energy spent on the server (by any system) is proportional to the change of temperature within one unit of time:

$$E_t = |\Delta T_t| = |T_{t+1} - T_t|$$

thus:

$E_t = T_{t+1} - T_t$ if $T_{t+1} > T_t$, that is, if the server is heated up,

$E_t = T_t - T_{t+1}$ if $T_{t+1} < T_t$, that is, if the server is cooled down,

then that means that the energy saved by the AI at each iteration t (each minute) is equal to the difference in absolute changes of temperatures caused in the server between the unintelligent server's integrated cooling system and the AI from t and $t+1$:

Energy saved by the AI between t and $t+1$

$$= \left| \Delta T_t^{\text{Server's Integrated Cooling System}} \right| - \left| \Delta T_t^{\text{AI}} \right|$$

$$= \left| \Delta T_t^{\text{noAI}} \right| - \left| \Delta T_t^{\text{AI}} \right|$$

where:

1. ΔT_t^{noAI} is the change of temperature that the server's integrated cooling system would cause in the server during the iteration t, that is, from t to $t+1$ minute.

2. ΔT_t^{AI} is the change of temperature that the AI would cause in the server during the iteration t, that is, from t to $t+1$ minute.

The AIs goal is to save as much as it can every minute, therefore saving the maximum total energy over 1 full year of simulation, and eventually saving the business the maximum cost possible on their cooling/heating electricity bill. That's how we do business in the 21st century; with AI!

Now that we fully understand how our server environment works, and how it's simulated, it's time to proceed with what absolutely must be done when defining an AI environment. You know the next steps already:

1. Defining the states.
2. Defining the actions.
3. Defining the rewards.

Defining the states

Remember, when you're doing deep Q-learning, the input state is always a 1D vector. (Unless you are doing deep convolutional Q-learning, in which case the input state is a 2D image, but that's getting ahead of ourselves! Wait for *Chapter 12, Deep Convolution Q-Learning*). So, what will the input state vector be in this server environment? What information will it contain in order to describe well enough each state of the environment? These are the questions you must ask yourself when modeling an AI problem and building the environment. Try to answer these questions first on your own and figure out the input state vector in this case, and you can find out what we're using in the next paragraph. Hint: have a look again at the variable defined preceding.

The input state s_t at time t is composed of the following three elements:

1. The temperature of the server at time t
2. The number of users in the server at time t
3. The rate of data transmission in the server at time t

Thus, the input state will be an input vector of these three elements. Our future AI will take this vector as input, and will return an action to perform at each time, t. Speaking of the actions, what are they going to be? Let's find out.

Defining the actions

To figure out which actions to perform, we need to remember the goal, which is to optimally regulate the temperature of the server. The actions are simply going to be the temperature changes that the AI can cause inside the server, in order to heat it up or cool it down. In deep Q-learning, the actions must always be discrete; they can't be plucked from a range, we need a defined number of possible actions. Therefore, we'll consider five possible temperature changes, from $-3°C$ to $+3°C$, so that we end up with five possible actions that the AI can perform to regulate the temperature of the server:

Action	What it does
0	The AI cools down the server by 3°C
1	The AI cools down the server by 1.5°C
2	The AI does not transfer any heat to the server (no temperature change)
3	The AI heats up the server by 1.5°C
4	The AI heats up the server by 3°C

Figure 1: Defining the actions

Great. Finally, let's see how we're going to reward and punish our AI.

Defining the rewards

You might have guessed from the earlier *Overall functioning* section what the reward is going to be. The reward at iteration t is the energy saved by the AI, with respect to how much energy the server's integrated cooling system would have spent; that is, the difference between the energy that the unintelligent cooling system would spend if the AI was deactivated, and the energy that the AI spends on the server:

$$\text{Reward}_t = E_t^{\text{noAI}} - E_t^{\text{AI}}$$

Since according to *Assumption 2*, the energy spent is equal to the change of the temperature induced in the server (by any system, including the AI or the unintelligent cooling system):

$$E_t = |\Delta T_t| = |T_{t+1} - T_t|$$

thus:

$E_t = T_{t+1} - T_t$ if $T_{t+1} > T_t$, if the server is heated up,

$E_t = T_t - T_{t+1}$ if $T_{t+1} < T_t$, if the server is cooled down,

then we receive a reward at time t that is the difference in the change of temperature caused in the server between unintelligent cooling system (that is when there is no AI) and the AI:

Energy saved by the AI between t and $t+1$

$$= E_t^{\text{noAI}} - E_t^{\text{AI}}$$

$$= \left|\Delta T_t^{\text{noAI}}\right| - \left|\Delta T_t^{\text{AI}}\right|$$

where:

1. ΔT_t^{noAI} is the change of temperature that the server's integrated cooling system would cause in the server during the iteration t, that is, from t to $t+1$ minute.

2. ΔT_t^{AI} is the change of temperature that the AI would cause in the server during the iteration t, that is, from t to $t+1$ minute.

Important note: It's important to understand that the systems (our AI and the server's integrated cooling system) will be evaluated separately, in order to compute the rewards. Since at each time point the actions of the two different systems lead to different temperatures, we have to keep track of the two temperatures separately, as T_t^{AI} and T_t^{noAI}. In other words, we're performing two separate simulations at the same time, following the same fluctuations of users and data; one for the AI, and one for the server's integrated cooling system.

To complete this section, we'll do a small simulation of 2 iterations (that is, 2 minutes) as an example to make everything crystal clear.

Final simulation example

Let's say that we're at time $t = 4:00$ pm, and that the temperature of the server is $T_t = 28°C$, both with the AI and without it. At this exact time, the AI predicts an action: 0, 1, 2, 3 or 4. Since, right now, the server's temperature is outside the optimal temperature range, $\left[18°C, 24°C \right]$, the AI will probably predict actions 0, 1 or 2. Let's say that it predicts 1, which corresponds to cooling the server down by $1.5°C$. Therefore, between $t = 4:00$ pm and $t+1 = 4:01$ pm, the AI makes the server's temperature go from $T_t^{\text{AI}} = 28°C$ to $T_{t+1}^{\text{AI}} = 26.5°C$:

$$\Delta T_t^{\text{AI}}$$

$$= T_{t+1}^{\text{AI}} - T_t^{\text{AI}}$$

$$= 26.5 - 28$$

$$= -1.5°C$$

Thus, based on *Assumption 2*, the energy spent by the AI on the server is:

$$E_t^{\text{AI}}$$

$$= \left| \Delta T_t^{\text{AI}} \right|$$

$$= 1.5 \, \text{Joules}$$

Now only one piece of information is missing to compute the reward: the energy that the server's integrated cooling system would have spent if the AI was deactivated between 4:00 pm and 4:01 pm. Remember that this unintelligent cooling system automatically brings the server's temperature back to the closest bound of the optimal temperature range $\left[18°\text{C}, 24°\text{C} \right]$. Since at $t = 4:00$ pm the temperature was $T_t = 28°\text{C}$, then the closest bound of the optimal temperature range at that time was $24°\text{C}$. Thus, the server's integrated cooling system would have changed the temperature from $T_t = 28°\text{C}$ to $T_{t+1} = 24°\text{C}$, and the server's temperature change that would have occurred if there was no AI is:

$$\Delta T_t^{\text{noAI}}$$

$$= T_{t+1}^{\text{noAI}} - T_t^{\text{noAI}}$$

$$= 24 - 28$$

$$= -4°C$$

Based on *Assumption 2*, the energy that the unintelligent cooling system would have spent if there was no AI is:

$$E_t^{\text{noAI}}$$

$$= \left| \Delta T_t^{\text{noAI}} \right|$$

$$= 4 \, \text{Joules}$$

In conclusion, the reward the AI gets after playing this action at time $t = 4:00$ pm is:

$$\text{Reward}$$

$$= E_t^{\text{noAI}} - E_t^{\text{AI}}$$

$$= 4 - 1.5$$

$$= 2.5$$

I'm sure you'll have noticed that as it stands, our AI system doesn't involve itself with the optimal range of temperatures for the server; as I've mentioned before, everything comes from the rewards, and the AI doesn't get any reward for being inside the optimal range or any penalty for being outside it. Once we've built the AI completely, I recommend that you play around with the code and try adding some rewards or penalties that get the AI to stick close to the optimal range; but for now, to keep things simple and get our AI up and running, we'll leave the reward as entirely linked to energy saved.

Then, between $t = 4:00$ pm and $t+1 = 4:01$ pm, new things happen: some new users log on to the server, some existing users log off, some new data transmits into the server, and some existing data transmits out. Based on *Assumption 1*, these factors make the server's temperature change. Let's say that overall, they increase the server's temperature by 5°C:

$$\Delta_t \text{ Intrinsic Temperature} = 5°C$$

Now, remember that we're evaluating two systems separately: our AI, and the server's integrated cooling system. Therefore we must compute the two temperatures we would get with each of these two systems separately, one without the other, at $t+1 = 4:01$ pm. Let's start with the AI.

The temperature we get at $t+1 = 4:01$ pm when the AI is activated is:

$$T_{t+1}^{AI}$$

$$= T_t^{AI} + \Delta T_t^{AI} + \Delta_t \text{ Intrinsic Temperature}$$

$$= 28 + (-1.5) + 5$$

$$= 31.5°C$$

And the temperature we get at $t+1 = 4:01$ pm if the AI is not activated is:

$$T_{t+1}^{noAI}$$

$$= T_t^{noAI} + \Delta T_t^{noAI} + \Delta_t \text{ Intrinsic Temperature}$$

$$= 28 + (-4) + 5$$

$$= 29°C$$

Now we have our two separate temperatures, which are T^{AI}_{t+1} = 31.5°C when the AI is activated, and T^{noAI}_{t+1} = 29°C when the AI is not activated.

Let's simulate what happens between $t+1 = 4:01$ pm and $t+2 = 4:02$ pm. Again, our AI will make a prediction, and since the server is heating up, let's say it predicts action 0, which corresponds to cooling down the server by $3°C$, bringing it down to $T^{AI}_{t+2} = 28.5°C$. Therefore, the energy spent by the AI between $t+1 = 4:01$ pm and $t+2 = 4:02$ pm is:

$$E^{AI}_{t+1}$$

$$= \left| \Delta T^{AI}_{t+1} \right|$$

$$= \left| 28.5 - 31.5 \right|$$

$$= 3 \, \text{Joules}$$

Now regarding the server's integrated cooling system (that is, when there is no AI), since at $t+1 = 4:01$ pm we had $T^{noAI}_{t+1} = 29°C$, then the closest bound of the optimal range of temperatures is still $24°C$, and so the energy that the server's unintelligent cooling system would spend between $t+1 = 4:01$ pm and $t+2 = 4:02$ pm is:

$$E^{noAI}_{t+1}$$

$$= \left| \Delta T^{noAI}_{t+1} \right|$$

$$= \left| 24 - 29 \right|$$

$$= 5 \, \text{Joules}$$

Hence the reward obtained between $t+1 = 4:01$ pm and $t+2 = 4:02$ pm, which is only and entirely based on the amount of energy saved, is:

$$\text{Reward}$$

$$= E^{noAI}_{t+1} - E^{AI}_{t+1}$$

$$= 5 - 3$$

$$= 2$$

Finally, the total reward obtained between $t = 4:00$ pm and $t+2 = 4:02$ pm is:

$$\text{Total Reward}$$

$$= \left(\text{Reward obtained between t and t+1}\right) + \left(\text{Reward obtained between t+1 and t+2}\right)$$

$$= 2.5 + 2$$

$$= 4.5$$

That was an example of the whole process happening for two minutes. In our implementation we'll run the same process over 1000 epochs of 5-month periods for the training, and then, once our AI is trained, we'll run the same process over 1 full year of simulation for the testing.

Now that we've defined and built the environment in detail, it's time for our AI to take action! This is where deep Q-learning comes into play. Our model will be more advanced than the previous one because I'm introducing some new tricks, called **dropout** and **early stopping**, which are great techniques for you to have in your toolkit; they usually improve the training performance of deep Q-learning.

Don't forget, you'll also get an AI Blueprint, which will allow you to adapt what we do here to any other business problem that you want to solve with deep Q-learning.

Ready? Let's smash this.

AI solution

Let's start by reminding ourselves of the whole deep Q-learning model, while adapting it to this case study, so that you don't have to scroll or turn many pages back into the previous chapters. Repetition is never bad; it sticks the knowledge into our heads more firmly. Here's the deep Q-learning algorithm for you again:

Initialization:

1. The memory of the experience replay is initialized to an empty list, called `memory` in the code (the `dqn.py` Python file in the `Chapter 11` folder of the GitHub repo).
2. We choose a maximum size for the memory, called `max_memory` in the code (the `dqn.py` Python file in the `Chapter 11` folder of the GitHub repo).

At each time *t* (each minute), we repeat the following process, until the end of the epoch:

1. We predict the Q-values of the current state S_t. Since five actions can be performed (0 == Cooling 3°C, 1 == Cooling 1.5°C, 2 == No Heat Transfer, 3 == Heating 1.5°C, 4 == Heating 3°C), we get five predicted Q-values.

2. We perform the action selected by the argmax method, which simply consists of selecting the action that has the highest of the five predicted Q-values:

$$a_t = \underset{a}{\operatorname{argmax}}\left\{Q(s_t, a)\right\}$$

3. We get the reward $R(s_t, a_t)$, which is the difference $E_t^{\text{noAI}} - E_t^{\text{AI}}$.

4. We reach the next state s_{t+1}, which is composed of the three following elements:

    ○   The temperature of the server at time $t+1$

    ○   The number of users in the server at time $t+1$

    ○   The rate of data transmission in the server at time $t+1$

5. We append the transition (s_t, a_t, r_t, s_{t+1}) in the memory.

6. We take a random batch $B \subset M$ of transitions. For all the transitions $\left(s_{t_B}, a_{t_B}, r_{t_B}, s_{t_B+1}\right)$ of the random batch B:

    ○   We get the predictions: $Q\left(s_{t_B}, a_{t_B}\right)$

    ○   We get the targets: $R(s_{t_B}, a_{t_B}) + \gamma \underset{a}{\max}\left(Q(s_{t_B+1}, a)\right)$

    ○   We compute the loss between the predictions and the targets over the whole batch B:

$$\text{Loss} = \frac{1}{2}\sum_B \left(R(s_{t_B}, a_{t_B}) + \gamma \underset{a}{\max}(Q(s_{t_B+1}, a)) - Q(s_{t_B}, a_{t_B})\right)^2 = \frac{1}{2}\sum_B TD_{t_B}(s_{t_B}, a_{t_B})^2$$

And then finally we backpropagate this loss error back into the neural network, and through stochastic gradient descent we update the weights according to how much they contributed to the loss error.

I hope the refresher was refreshing! Let's move on to the brain of the outfit.

The brain

By the brain, I mean of course the artificial neural network of our AI.

Our brain will be a fully connected neural network, composed of two hidden layers, the first one with 64 neurons, and the second one with 32 neurons. As a reminder, this neural network takes as inputs the states of the environment, and returns as outputs the Q-values for each of the five possible actions.

This particular design of a neural network, with two hidden layers of 64 and 32 neurons respectively, is considered something of a **classic** architecture. It's suitable to solve a lot of problems, and it will work well for us here.

This artificial brain will be trained with a **Mean Squared Error** (MSE) loss, and an Adam optimizer. The choice for the MSE loss is because we want to measure and reduce the squared difference between the predicted value and the target value, and the Adam optimizer is a classic optimizer used, in practice, by default.

Here is what this artificial brain looks like:

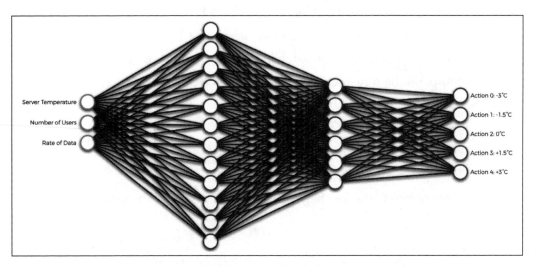

Figure 2: The artificial brain of our AI

This artificial brain looks complex to create, but we can build it very easily thanks to the amazing Keras library. In the last chapter, we used PyTorch because it's the neural network library I'm more familiar with; but I want you to be able to use as many AI tools as possible, so in this chapter we're going to power on with Keras. Here's a preview of the full implementation containing the part that builds this brain all by itself (taken from the brain_nodropout.py file):

```
# BUILDING THE BRAIN

class Brain(object):

    # BUILDING A FULLY CONNECTED NEURAL NETWORK DIRECTLY INSIDE THE
INIT METHOD

    def __init__(self, learning_rate = 0.001, number_actions = 5):
        self.learning_rate = learning_rate
```

```
# BUILDING THE INPUT LAYER COMPOSED OF THE INPUT STATE
states = Input(shape = (3,))

# BUILDING THE FULLY CONNECTED HIDDEN LAYERS
x = Dense(units = 64, activation = 'sigmoid')(states)
y = Dense(units = 32, activation = 'sigmoid')(x)

# BUILDING THE OUTPUT LAYER, FULLY CONNECTED TO THE LAST
HIDDEN LAYER
q_values = Dense(units = number_actions, activation =
'softmax')(y)

# ASSEMBLING THE FULL ARCHITECTURE INSIDE A MODEL OBJECT
self.model = Model(inputs = states, outputs = q_values)

# COMPILING THE MODEL WITH A MEAN-SQUARED ERROR LOSS AND A
CHOSEN OPTIMIZER
self.model.compile(loss = 'mse', optimizer = Adam(lr =
learning_rate))
```

As you can see, it only takes a couple of lines of code, and I'll explain every line of that code to you in a later section. Now let's move on to the implementation.

Implementation

This implementation will be divided into five parts, each part having its own Python file. You can find the full implementation in the `Chapter 11` folder of the GitHub repository. These five parts constitute the general AI framework, or AI Blueprint, that should be followed whenever you build an environment to solve any business problem with deep reinforcement learning.

Here they are, from Step 1 to Step 5:

- **Step 1**: Building the environment (`environment.py`)
- **Step 2**: Building the brain (`brain_nodropout.py` or `brain_dropout.py`)
- **Step 3**: Implementing the deep reinforcement learning algorithm, which in our case is a deep Q-learning model (`dqn.py`)
- **Step 4**: Training the AI (`training_noearlystopping.py` or `training_earlystopping.py`)
- **Step 5**: Testing the AI (`testing.py`)

In order, those are the main steps of the general AI framework.

We'll follow this AI Blueprint to implement the AI for our specific case in the following five sections, each corresponding to one of these five main steps. Within each step, we'll distinguish the sub-steps that are still part of the general AI framework from the sub-steps that are specific to our project by writing the titles of the code sections in capital letters for all the sub-steps of the general AI framework, and in lowercase letters for all the sub-steps specific to our project.

That means that anytime you see a new code section where the title is written in capital letters, then it is the next sub-step of the general AI framework, which you should also follow when building an AI for your own business problem.

This next step, building the environment, is the largest Python implementation file for this project. Make sure you're rested and your batteries are recharged, and as soon as you are ready, let's tackle this together!

Step 1 – Building the environment

In this first step, we are going to build the environment inside a class. Why a class? Because we would like our environment to be an object which we can easily create with any values we choose for some parameters.

For example, we can create one environment object for a server that has a certain number of connected users and a certain rate of data at a specific time, and another environment object for a different server that has a different number of connected users and a different rate of data. Thanks to the advanced structure of this class, we can easily plug-and-play the environment objects we create on different servers which have their own parameters, regulating their temperatures with several different AIs, so that we can minimize the energy consumption of a whole data center, just as Google DeepMind did for Google's data centers with its DQN (deep Q-learning) algorithm.

This class follows the following sub-steps, which are part of the general AI Framework inside Step 1 – Building the environment:

- **Step 1-1**: Introducing and initializing all the parameters and variables of the environment.

- **Step 1-2**: Making a method that updates the environment right after the AI plays an action.

- **Step 1-3**: Making a method that resets the environment.

- **Step 1-4**: Making a method that gives us at any time the current state, the last reward obtained, and whether the game is over.

You'll find the whole implementation of this Environment class in this section. Remember the most important thing: all the code sections with their titles written in capital letters are steps of the general AI framework/Blueprint, and all the code sections having their titles written in lowercase letters are specific to our case study.

The implementation of the environment has 144 lines of code. I won't explain each line of code for two reasons:

1. It would make this chapter really overwhelming.

2. The code is very simple, is commented on for clarity, and just creates everything we've defined so far in this chapter.

I'm confident you'll have no problems understanding it. Besides, the code section titles and the chosen variable names are clear enough to understand the structure and the flow of the code at face value. I'll walk you through the code broadly. Here we go!

First, we start building the Environment class with its first method, the __init__ method, which introduces and initializes all the parameters and variables, as we described earlier:

```
# BUILDING THE ENVIRONMENT IN A CLASS

class Environment(object):

    # INTRODUCING AND INITIALIZING ALL THE PARAMETERS AND VARIABLES OF
THE ENVIRONMENT

    def __init__(self, optimal_temperature = (18.0, 24.0), initial_
month = 0, initial_number_users = 10, initial_rate_data = 60):
        self.monthly_atmospheric_temperatures = [1.0, 5.0, 7.0, 10.0,
11.0, 20.0, 23.0, 24.0, 22.0, 10.0, 5.0, 1.0]
        self.initial_month = initial_month
        self.atmospheric_temperature = self.monthly_atmospheric_
temperatures[initial_month]
        self.optimal_temperature = optimal_temperature
        self.min_temperature = -20
        self.max_temperature = 80
        self.min_number_users = 10
        self.max_number_users = 100
        self.max_update_users = 5
        self.min_rate_data = 20
        self.max_rate_data = 300
        self.max_update_data = 10
        self.initial_number_users = initial_number_users
```

```
        self.current_number_users = initial_number_users
        self.initial_rate_data = initial_rate_data
        self.current_rate_data = initial_rate_data
        self.intrinsic_temperature = self.atmospheric_temperature +
1.25 * self.current_number_users + 1.25 * self.current_rate_data
        self.temperature_ai = self.intrinsic_temperature
        self.temperature_noai = (self.optimal_temperature[0] + self.
optimal_temperature[1]) / 2.0
        self.total_energy_ai = 0.0
        self.total_energy_noai = 0.0
        self.reward = 0.0
        self.game_over = 0
        self.train = 1
```

You'll notice the `self.monthly_atmospheric_temperatures` variable; that's a list containing the average monthly atmospheric temperatures for each of the 12 months: 1°C in January, 5°C in February, 7°C in March, and so on.

The `self.atmospheric_temperature` variable is the current average atmospheric temperature of the month we're in during the simulation, and it's initialized as the atmospheric temperature of the initial month, which we'll set later as January.

The `self.game_over` variable tells the AI whether or not we should reset the temperature of the server, in case it goes outside the allowed range of [-20°C, 80°C]. If it does, `self.game_over` will be set equal to 1, otherwise it will remain at 0.

Finally, the `self.train` variable tells us whether we're in training mode or inference mode. If we're in training mode, `self.train = 1`. If we're in inference mode, `self.train = 0`. The rest is just putting into code everything we defined in words at the beginning of this chapter.

Let's move on!

Now, we make the second method, `update_env`, which updates the environment after the AI performs an action. This method takes three arguments as inputs:

1. `direction`: A variable describing the direction of the heat transfer the AI imposes on the server, like so: if `direction == 1`, the AI is heating up the server. If `direction == -1`, the AI is cooling down the server. We'll need to have the value of this direction before calling the `update_env` method, since this method is called after the action is performed.

2. `energy_ai`: The energy spent by the AI to heat up or cool down the server at this specific time when the action is played. Based on assumption 2, it will be equal to the temperature change caused by the AI in the server.

3. `month`: Simply the month we're in at the specific time when the action is played.

The first actions the program takes inside this method are to compute the reward. Indeed, right after the action is played, we can immediately deduce the reward, since it is the difference between the energy that the server's integrated system would spend if there was no AI, and the energy spent by the AI:

```
# MAKING A METHOD THAT UPDATES THE ENVIRONMENT RIGHT AFTER THE AI
PLAYS AN ACTION

def update_env(self, direction, energy_ai, month):

    # GETTING THE REWARD

    # Computing the energy spent by the server's cooling system
when there is no AI
    energy_noai = 0
    if (self.temperature_noai < self.optimal_temperature[0]):
        energy_noai = self.optimal_temperature[0] - self.
temperature_noai
        self.temperature_noai = self.optimal_temperature[0]
    elif (self.temperature_noai > self.optimal_temperature[1]):
        energy_noai = self.temperature_noai - self.optimal_
temperature[1]
        self.temperature_noai = self.optimal_temperature[1]
    # Computing the Reward
    self.reward = energy_noai - energy_ai
    # Scaling the Reward
    self.reward = 1e-3 * self.reward
```

You have probably noticed that we choose to scale the reward at the end. In short, scaling is bringing the values (here the rewards) down into a short range. For example, normalization is a scaling technique where all the values are brought down into a range between 0 and 1. Another widely used scaling technique is standardization, which will be explained a bit later on.

Scaling is a common practice that is usually recommended in research papers when performing deep reinforcement learning, as it stabilizes training and improves the performance of the AI.

After getting the reward, we reach the next state. Remember that each state is composed of the following elements:

1. The temperature of the server at time *t*
2. The number of users in the server at time *t*
3. The rate of data transmission in the server at time *t*

So, as we reach the next state, we update each of these elements one by one, following the sub-steps highlighted as comments in this next code section:

```
# GETTING THE NEXT STATE

# Updating the atmospheric temperature
self.atmospheric_temperature = self.monthly_atmospheric_
temperatures[month]
# Updating the number of users
self.current_number_users += np.random.randint(-self.max_
update_users, self.max_update_users)
if (self.current_number_users > self.max_number_users):
    self.current_number_users = self.max_number_users
elif (self.current_number_users < self.min_number_users):
    self.current_number_users = self.min_number_users
# Updating the rate of data
self.current_rate_data += np.random.randint(-self.max_update_
data, self.max_update_data)
if (self.current_rate_data > self.max_rate_data):
    self.current_rate_data = self.max_rate_data
elif (self.current_rate_data < self.min_rate_data):
    self.current_rate_data = self.min_rate_data
# Computing the Delta of Intrinsic Temperature
past_intrinsic_temperature = self.intrinsic_temperature
self.intrinsic_temperature = self.atmospheric_temperature +
1.25 * self.current_number_users + 1.25 * self.current_rate_data
delta_intrinsic_temperature = self.intrinsic_temperature -
past_intrinsic_temperature
# Computing the Delta of Temperature caused by the AI
if (direction == -1):
    delta_temperature_ai = -energy_ai
```

```
    elif (direction == 1):
        delta_temperature_ai = energy_ai
    # Updating the new Server's Temperature when there is the AI
        self.temperature_ai += delta_intrinsic_temperature + delta_
temperature_ai
        # Updating the new Server's Temperature when there is no AI
        self.temperature_noai += delta_intrinsic_temperature
```

Then, we update the `self.game_over` variable if needed, that is, if the temperature of the server goes outside the allowed range of [-20°C, 80°C]. This can happen if the server temperature goes below the minimum temperature of -20°C, or if the server temperature goes higher than the maximum temperature of 80°C. Plus we do two extra things: we bring the server temperature back into the optimal temperature range (closest bound), and since doing this spends some energy, we update the total energy spent by the AI (`self.total_energy_ai`). That's exactly what is coded in the next code section:

```
    # GETTING GAME OVER

    if (self.temperature_ai < self.min_temperature):
        if (self.train == 1):
            self.game_over = 1
        else:
            self.total_energy_ai += self.optimal_temperature[0] -
self.temperature_ai
            self.temperature_ai = self.optimal_temperature[0]
    elif (self.temperature_ai > self.max_temperature):
        if (self.train == 1):
            self.game_over = 1
        else:
            self.total_energy_ai += self.temperature_ai - self.
optimal_temperature[1]
            self.temperature_ai = self.optimal_temperature[1]
```

Now, I know it seems unrealistic for the server to snap right back to 24 degrees from 80, or to 18 from -20, but this is an action the magically efficient integrated cooling system we defined earlier is perfectly capable of. Think of it as the AI switching to the integrated system for a moment in the case of a temperature disaster. Once again, this is an area that will benefit enormously from your ongoing tinkering once we've got the AI up and running; after that, you can play around with these figures as you like in the interests of a more realistic server model.

Then, we update the two scores coming from the two separate simulations, which are:

1. `self.total_energy_ai`: The total energy spent by the AI
2. `self.total_energy_noai`: The total energy spent by the server's integrated cooling system when there is no AI.

```
# UPDATING THE SCORES

# Updating the Total Energy spent by the AI
self.total_energy_ai += energy_ai
# Updating the Total Energy spent by the server's cooling
system when there is no AI
self.total_energy_noai += energy_noai
```

Then to improve the performance, we scale the next state by scaling each of its three elements (server temperature, number of users, and data transmission rate). To do so, we perform a simple standardization scaling technique, which simply consists of subtracting the minimum value of the variable, and then dividing by the maximum delta of the variable:

```
# SCALING THE NEXT STATE

scaled_temperature_ai = (self.temperature_ai - self.min_
temperature) / (self.max_temperature - self.min_temperature)
scaled_number_users = (self.current_number_users - self.min_
number_users) / (self.max_number_users - self.min_number_users)
scaled_rate_data = (self.current_rate_data - self.min_rate_
data) / (self.max_rate_data - self.min_rate_data)
next_state = np.matrix([scaled_temperature_ai, scaled_number_
users, scaled_rate_data])
```

Finally, we end this `update_env` method by returning the next state, the reward received, and whether the game is over or not:

```
# RETURNING THE NEXT STATE, THE REWARD, AND GAME OVER

return next_state, self.reward, self.game_over
```

Great! We're done with this long, but important, method that updates the environment at each time step (each minute). Now there are two final and very easy methods to go: one that resets the environment, and one that gives us three pieces of information at any time: the current state, the last reward received, and whether or not the game is over.

Here's the `reset` method, which resets the environment when a new training episode starts, by resetting all the variables of the environment to their originally initialized values:

```
# MAKING A METHOD THAT RESETS THE ENVIRONMENT

    def reset(self, new_month):
        self.atmospheric_temperature = self.monthly_atmospheric_
temperatures[new_month]
        self.initial_month = new_month
        self.current_number_users = self.initial_number_users
        self.current_rate_data = self.initial_rate_data
        self.intrinsic_temperature = self.atmospheric_temperature +
1.25 * self.current_number_users + 1.25 * self.current_rate_data
        self.temperature_ai = self.intrinsic_temperature
        self.temperature_noai = (self.optimal_temperature[0] + self.
optimal_temperature[1]) / 2.0
        self.total_energy_ai = 0.0
        self.total_energy_noai = 0.0
        self.reward = 0.0
        self.game_over = 0
        self.train = 1
```

Finally, here's the `observe` method, which lets us know at any given time the current state, the last reward received, and whether the game is over:

```
# MAKING A METHOD THAT GIVES US AT ANY TIME THE CURRENT STATE, THE
LAST REWARD AND WHETHER THE GAME IS OVER

    def observe(self):
        scaled_temperature_ai = (self.temperature_ai - self.min_
temperature) / (self.max_temperature - self.min_temperature)
        scaled_number_users = (self.current_number_users - self.min_
number_users) / (self.max_number_users - self.min_number_users)
        scaled_rate_data = (self.current_rate_data - self.min_rate_
data) / (self.max_rate_data - self.min_rate_data)
        current_state = np.matrix([scaled_temperature_ai, scaled_
number_users, scaled_rate_data])
        return current_state, self.reward, self.game_over
```

Awesome! We're done with the first step of the implementation, building the environment. Now let's move on to the next step and start building the brain.

Step 2 – Building the brain

In this step, we're going to build the artificial brain of our AI, which is nothing other than a fully connected neural network. Here it is again:

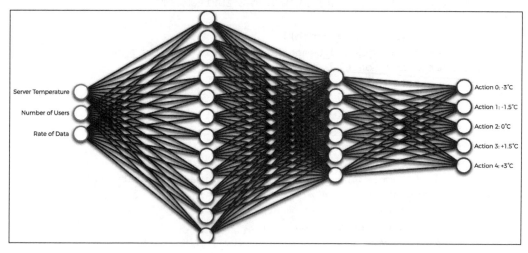

Figure 3: The artificial brain of our AI

We'll build this artificial brain inside a class for the same reason as before, which is to allow us to create several artificial brains, for different servers inside a data center. Maybe some servers will need different artificial brains with different hyper-parameters than other servers. That's why, thanks to this class/object advanced Python structure, we can easily switch from one brain to another, to regulate the temperature of a new server that requires an AI with different neural network parameters. That's the beauty of **Object-Oriented Programming (OOP)**.

We're building this artificial brain with the amazing Keras library. From this library, we use the Dense() class to create our two fully connected hidden layers, the first one from 64 hidden neurons, and the second one from 32 neurons. Remember, this is a classic neural network architecture often used by default, as common practice, and seen in many research papers. At the end, we use the Dense() class again to return the Q-values, which are the outputs of the artificial neural network.

Later on, when we code the training and testing files, we'll use the argmax method to select the action that has the maximum Q-value. Then, we assemble all the components of the brain, including the inputs and outputs, by creating it as an object of the Model() class (which is very useful in that we can save and load a model with specific weights). Finally, we'll compile it with a mean squared error loss and an Adam optimizer. I'll explain all this in more detail later.

Here are the new steps of the general AI framework:

- **Step 2-1**: Build the input layer, composed of the input states.
- **Step 2-2**: Build a defined number of hidden layers with a defined number of neurons inside each layer, fully connected to the input layer and between each other.
- **Step 2-3**: Build the output layer, fully connected to the last hidden layer.
- **Step 2-4**: Assemble the full architecture inside a model object.
- **Step 2-5**: Compile the model with a mean squared error loss function and a chosen optimizer.

The implementation of this is presented to you in a choice of two different files:

1. `brain_nodropout.py`: An implementation file that builds the artificial brain without the dropout regularization technique (I'll explain what it is very soon).
2. `brain_dropout.py`: An implementation file that builds the artificial brain with the dropout regularization technique.

First let me give you the implementation without dropout, and then I'll provide one with dropout and explain it.

Without dropout

Here is the full implementation of the artificial brain, without any dropout regularization technique:

```
1      # AI for Business - Minimize cost with Deep Q-Learning
2      # Building the Brain without Dropout
3
4      # Importing the libraries
5      from keras.layers import Input, Dense
6      from keras.models import Model
7      from keras.optimizers import Adam
8
9      # BUILDING THE BRAIN
10
11     class Brain(object):
12
```

```
13          # BUILDING A FULLY CONNECTED NEURAL NETWORK DIRECTLY INSIDE
        THE INIT METHOD

14

15          def __init__(self, learning_rate = 0.001, number_actions =
        5):

16              self.learning_rate = learning_rate

17

18              # BUILDING THE INPUT LAYER COMPOSED OF THE INPUT STATE

19              states = Input(shape = (3,))

20

21              # BUILDING THE FULLY CONNECTED HIDDEN LAYERS

22              x = Dense(units = 64, activation = 'sigmoid')(states)

23              y = Dense(units = 32, activation = 'sigmoid')(x)

24

25              # BUILDING THE OUTPUT LAYER, FULLY CONNECTED TO THE LAST
        HIDDEN LAYER

26              q_values = Dense(units = number_actions, activation =
        'softmax')(y)

27

28              # ASSEMBLING THE FULL ARCHITECTURE INSIDE A MODEL OBJECT

29              self.model = Model(inputs = states, outputs = q_values)

30

31              # COMPILING THE MODEL WITH A MEAN-SQUARED ERROR LOSS AND
        A CHOSEN OPTIMIZER

32              self.model.compile(loss = 'mse', optimizer = Adam(lr =
        learning_rate))
```

Now, let's go through the code in detail.

Line 5: We import the `Input` and `Dense` classes from the `layers` module in the `keras` library. The `Input` class allows us to build the input layer, and the `Dense` class allows us to build the fully-connected layers.

Line 6: We import the `Model` class from the `models` module in the `keras` library. It allows us to build the whole neural network model by assembling its different layers.

Line 7: We import the Adam class from the optimizers module in the keras library. It allows us to use the Adam optimizer, used to update the weights of the neural network through stochastic gradient descent, when backpropagating the loss error in each iteration of the training.

Line 11: We introduce the Brain class, which will contain not only the whole architecture of the artificial neural network, but also the connection of the model to the loss (Mean-Squared Error) and the Adam optimizer.

Line 15: We introduce the __init__ method, which will be the only method of this class. We define the whole architecture of the neural network inside it, just by creating successive variables which together assemble the neural network. This method takes as inputs two arguments:

1. The learning rate (learning_rate), which is a measure of how fast you want the neural network to learn (the higher the learning rate, the faster the neural network learns; but at the cost of quality). The default value is 0.001.

2. The number of actions (number_actions), which is of course the number of actions that our AI can perform. Now you might be thinking: why do we need to put that as an argument? Well that's just in case you want to build another AI that can perform more or fewer actions. In which case you would simply need to change the value of the argument and that's it. Pretty practical, isn't it?

Line 16: We create an object variable for the learning rate, self.learning_rate, initialized as the value of the learning_rate argument provided in the __init__ method (therefore the argument of the Brain class when we create the object in the future).

Line 19: We create the input states layer, called states, as an object of the Input class. Into this Input class we enter one argument, shape = (3,), which simply tells that the input layer is a 1D vector composed of three elements (the server temperature, the number of users, and the data transmission rate).

Line 22: We create the first fully-connected hidden layer, called x, as an object of the Dense class, which takes as input two arguments:

1. units: The number of hidden neurons we want to have in this first hidden layer. Here, we choose to have 64 hidden neurons.

2. `activation`: The activation function used to pass on the signal when forward-propagating the inputs into this first hidden layer. Here we choose, by default, a sigmoid activation function, which is as follows:

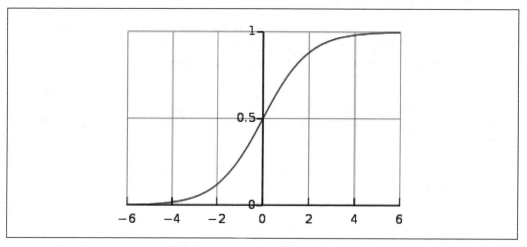

Figure 4: The sigmoid activation function

The ReLU activation function would also have worked well here; I encourage you to experiment! Note also how the connection from the input layer to this first hidden layer is made by calling the `states` variable right after the `Dense` class.

Line 23: We create the second fully-connected hidden layer, called `y`, as an object of the `Dense` class, which takes as input the same two arguments:

1. `units`: The number of hidden neurons we want to have in this second hidden layer. This time we choose to have 32 hidden neurons.

2. `activation`: The activation function used to pass on the signal when forward-propagating the inputs into this first hidden layer. Here, again, we choose a sigmoid activation function.

Note once again how the connection from the first hidden layer to this second hidden layer is made by calling the `x` variable right after the `Dense` class.

Line 26: We create the output layer, called `q_values`, fully connected to the second hidden layer, as an object of the `Dense` class. This time, we input `number_actions` units since the output layer contains the actions to play, and a `softmax` activation function, as seen in *Chapter 5, Your First AI Model – Beware the Bandits!*, on the deep Q-learning theory.

Line 29: Using the `Model` class, we assemble the successive layers of the neural network, by just inputting the `states` as the inputs, and the `q_values` as the outputs.

Line 32: Using the `compile` method taken from the `Model` class, we connect our model to the Mean-Squared Error loss and the Adam optimizer. The latter takes the `learning_rate` argument as input.

With dropout

It'll be valuable for you to add one more powerful technique to your toolkit: **dropout**.

Dropout is a regularization technique that prevents overfitting, which is the situation where the AI model performs well on the training set, but poorly on the test set. Dropout simply consists of deactivating a randomly selected portion of neurons during each step of forward- and back-propagation. That means not all the neurons learn the same way, which prevents the neural network from overfitting the training data.

Adding dropout is very easy with `keras`. You simply need to call the `Dropout` class right after the `Dense` class, and input the proportion of neurons you want to deactivate, like so:

```
# AI for Business - Minimize cost with Deep Q-Learning
# Building the Brain with Dropout

# Importing the libraries
from keras.layers import Input, Dense, Dropout
from keras.models import Model
from keras.optimizers import Adam

# BUILDING THE BRAIN

class Brain(object):

    # BUILDING A FULLY CONNECTED NEURAL NETWORK DIRECTLY INSIDE THE
INIT METHOD

    def __init__(self, learning_rate = 0.001, number_actions = 5):
        self.learning_rate = learning_rate

        # BUILDING THE INPUT LAYER COMPOSED OF THE INPUT STATE
        states = Input(shape = (3,))

        # BUILDING THE FIRST FULLY CONNECTED HIDDEN LAYER WITH DROPOUT
    ACTIVATED
        x = Dense(units = 64, activation = 'sigmoid')(states)
        x = Dropout(rate = 0.1)(x)
```

```
        # BUILDING THE SECOND FULLY CONNECTED HIDDEN LAYER WITH
DROPOUT ACTIVATED
        y = Dense(units = 32, activation = 'sigmoid')(x)
        y = Dropout(rate = 0.1)(y)

        # BUILDING THE OUTPUT LAYER, FULLY CONNECTED TO THE LAST
HIDDEN LAYER
        q_values = Dense(units = number_actions, activation =
'softmax')(y)

        # ASSEMBLING THE FULL ARCHITECTURE INSIDE A MODEL OBJECT
        self.model = Model(inputs = states, outputs = q_values)

        # COMPILING THE MODEL WITH A MEAN-SQUARED ERROR LOSS AND A
CHOSEN OPTIMIZER
        self.model.compile(loss = 'mse', optimizer = Adam(lr =
learning_rate))
```

Here, we apply dropout to the first and second fully-connected layers, by deactivating 10% of their neurons each. Now, let's move on to the next step of our general AI framework: Step 3 – Implementing the deep reinforcement learning algorithm.

Step 3 – Implementing the deep reinforcement learning algorithm

In this new implementation (given in the dqn.py file), we simply have to follow the deep Q-learning algorithm provided before. Hence, this implementation follows the following sub-steps, which are part of the general AI framework:

- **Step 3-1**: Introduce and initialize all the parameters and variables of the deep Q-learning model.
- **Step 3-2**: Make a method that builds the memory in experience replay.
- **Step 3-3**: Make a method that builds and returns two batches of 10 inputs and 10 targets.

First, have a look at the whole code, and then I'll explain it line by line:

```
1       # AI for Business - Minimize cost with Deep Q-Learning
2       # Implementing Deep Q-Learning with Experience Replay
3
4       # Importing the libraries
```

```
5      import numpy as np
6

7      # IMPLEMENTING DEEP Q-LEARNING WITH EXPERIENCE REPLAY
8

9      class DQN(object):
10

11         # INTRODUCING AND INITIALIZING ALL THE PARAMETERS AND
       VARIABLES OF THE DQN
12         def __init__(self, max_memory = 100, discount = 0.9):
13             self.memory = list()
14             self.max_memory = max_memory
15             self.discount = discount
16

17         # MAKING A METHOD THAT BUILDS THE MEMORY IN EXPERIENCE
       REPLAY
18         def remember(self, transition, game_over):
19             self.memory.append([transition, game_over])
20             if len(self.memory) > self.max_memory:
21                 del self.memory[0]
22

23         # MAKING A METHOD THAT BUILDS TWO BATCHES OF INPUTS AND
       TARGETS BY EXTRACTING TRANSITIONS FROM THE MEMORY
24         def get_batch(self, model, batch_size = 10):
25             len_memory = len(self.memory)
26             num_inputs = self.memory[0][0][0].shape[1]
27             num_outputs = model.output_shape[-1]
28             inputs = np.zeros((min(len_memory, batch_size), num_
       inputs))
29             targets = np.zeros((min(len_memory, batch_size), num_
       outputs))
30             for i, idx in enumerate(np.random.randint(0, len_memory,
       size = min(len_memory, batch_size))):
31                 current_state, action, reward, next_state = self.
       memory[idx][0]
32                 game_over = self.memory[idx][1]
33                 inputs[i] = current_state
34                 targets[i] = model.predict(current_state)[0]
35                 Q_sa = np.max(model.predict(next_state)[0])
```

```
36                    if game_over:
37                        targets[i, action] = reward
38                    else:
39                        targets[i, action] = reward + self.discount *
        Q_sa
40            return inputs, targets
```

Line 5: We import the `numpy` library, because we'll be working with `numpy` arrays.

Line 9: We introduce the `DQN` class (**DQN** stands for **Deep Q-Network**), which contains the main parts of the deep Q-Learning algorithm, including experience replay.

Line 12: We introduce the `__init__` method, which creates the three following object variables of the `DQN` model: the experience replay memory, the capacity (maximum size of the memory), and the discount factor in the formula of the target. It takes as arguments `max_memory` (the capacity) and `discount` (the discount factor), in case we want to build other experience replay memories with different capacities, or if we want to change the value of the discount factor in the computation of the target. The default values of these arguments are respectively `100` and `0.9`, which were chosen arbitrarily and turned out to work quite well; these are good arguments to experiment with, to see what difference it makes when you set them differently.

Line 13: We create the experience replay memory object variable, `self.memory`, and we initialize it as an empty list.

Line 14: We create the object variable for the memory capacity, `self.max_memory`, and we initialize it as the value of the `max_memory` argument.

Line 15: We create the object variable for the discount factor, `self.discount`, and we initialize it as the value of the `discount` argument.

Line 18: We introduce the `remember` method, which takes as input a transition to be added to the memory, and `game_over`, which states whether or not this transition leads the server's temperature to go outside of the allowed range of temperatures.

Line 19: Using the `append` function called from the `memory` list, we add the transition with the `game_over` boolean into the memory (in the last position).

Line 20: If, after adding this transition, the size of the memory exceeds the memory capacity (`self.max_memory`).

Line 21: We delete the first element of the memory.

Line 24: We introduce the `get_batch` method, which takes as inputs the model we built in the previous Python file (`model`) and a batch size (`batch_size`), and builds two batches of inputs and targets by extracting 10 transitions from the memory (if the batch size is 10).

Line 25: We get the current number of elements in the memory and put it into a new variable, `len_memory`.

Line 26: We get the number of elements in the input state vector (which is 3), but instead of directly entering 3, we access this number from the `shape` attribute of the input state vector element of the memory, which we get by taking the `[0] [0] [0]` indexes. Each element of the memory is structured as follows:

`[[current_state, action, reward, next_state], game_over]`

Thus in `[0] [0] [0]`, the first `[0]` corresponds to the first element of the memory (meaning the first transition), the second `[0]` corresponds to the tuple [`current_state, action, reward, next_state`], and so the third `[0]` corresponds to the `current_state` element of that tuple. Hence, `self.memory[0] [0] [0]` corresponds to the first current state, and by adding `.shape[1]` we get the number of elements in that input state vector. You might be wondering why we didn't enter 3 directly; that's because we want to generalize this code to any input state vector dimension you might want to have in your environment. For example, you might want to consider an input state with more information about your server, such as the humidity. Thanks to this line of code, you won't have to change anything regarding your new number of state elements.

Line 27: We get the number of elements of the model output, meaning the number of actions. Just like on the previous line, instead of entering directly 5, we generalize by accessing this from the `shape` attribute called from our `model` object of the `Model` class. `-1` means that we get the last index of that `shape` attribute, where the number of actions is contained.

Line 28: We introduce and initialize the batch of inputs as a `numpy` array, of `batch_size` = 10 rows and 3 columns corresponding to input state elements, with only zeros. If the memory doesn't have 10 transitions yet, the number of rows will just be the length of the memory.

If the memory already has at least 10 transitions, what we get with this line of code is the following:

	Number of Users	Rate of Data	Server Temperature
Input State 1	0	0	0
Input State 2	0	0	0
Input State 3	0	0	0
Input State 4	0	0	0
Input State 5	0	0	0
Input State 6	0	0	0
Input State 7	0	0	0
Input State 8	0	0	0
Input State 9	0	0	0
Input State 10	0	0	0

Figure 5: Batch of inputs (1/2)

Line 29: We introduce and initialize the batch of targets as a numpy array of batch_size = 10 rows and 5 columns corresponding to the five possible actions, with only zeros. Just like before, if the memory doesn't have 10 transitions yet, the number of rows will just be the length of the memory. If the memory already has at least 10 transitions, what we get with this line of code is the following:

	Action 0	Action 1	Action 2	Action 3	Action 4
Target 1	0	0	0	0	0
Target 2	0	0	0	0	0
Target 3	0	0	0	0	0
Target 4	0	0	0	0	0
Target 5	0	0	0	0	0
Target 6	0	0	0	0	0
Target 7	0	0	0	0	0
Target 8	0	0	0	0	0
Target 9	0	0	0	0	0
Target 10	0	0	0	0	0

Figure 6: Batch of targets (1/3)

Line 30: We do a double iteration inside the same for loop. The first iterative variable i goes from 0 to the batch size (or up to len_memory if len_memory < batch_size):

$$i = 0, 1, 2, 3, 4, 5, 6, 7, 8, 9$$

That way, i will iterate each element of the batch. The second iterative variable idx takes 10 random indexes of the memory, in order to extract 10 random transitions from the memory. Inside the for loop, we populate the two batches of inputs and targets with their right values by iterating through each of their elements.

Line 31: We get the transition of the sampled index `idx` from the memory, composed of the current state, the action, the reward, and the next state. The reason we add `[0]` is because an element of the memory is structured as follows:

`[[current_state, action, reward, next_state], game_over]`

We'll get the `game_over` value separately, in the next line of code.

Line 32: We get the `game_over` value corresponding to that same index `idx` of the memory. As you can see, this time we add `[1]` on the end to get the second element of a memory element:

`[[current_state, action, reward, next_state], game_over]`

Line 33: We populate the batch of inputs with all the current states, in order to get this at the end of the `for` loop:

	Number of Users	Rate of Data	Server Temperature
Input State 1	U1	D1	T1
Input State 2	U2	D2	T2
Input State 3	U3	D3	T3
Input State 4	U4	D4	T4
Input State 5	U5	D5	T5
Input State 6	U6	D6	T6
Input State 7	U7	D7	T7
Input State 8	U8	D8	T8
Input State 9	U9	D9	T9
Input State 10	U10	D10	T10

Figure 7: Batch of inputs (2/2)

Line 34: Now we start populating the batch of targets with the right values. First, we populate it with all the Q-values $Q\left(s_{t_B}, a_{t_B}\right)$ that the model predicts for the different state-action pairs: (current state, action 0), (current state, action 1), (current state, action 2), (current state, action 3), and (current state, action 4). Thus we first get this (at the end of the `for` loop):

	Action 0	Action 1	Action 2	Action 3	Action 4
Target 1	Q(State 1, Action 0)	Q(State 1, Action 1)	Q(State 1, Action 2)	Q(State 1, Action 3)	Q(State 1, Action 4)
Target 2	Q(State 2, Action 0)	Q(State 2, Action 1)	Q(State 2, Action 2)	Q(State 2, Action 3)	Q(State 2, Action 4)
Target 3	Q(State 3, Action 0)	Q(State 3, Action 1)	Q(State 3, Action 2)	Q(State 3, Action 3)	Q(State 3, Action 4)
Target 4	Q(State 4, Action 0)	Q(State 4, Action 1)	Q(State 4, Action 2)	Q(State 4, Action 3)	Q(State 4, Action 4)
Target 5	Q(State 5, Action 0)	Q(State 5, Action 1)	Q(State 5, Action 2)	Q(State 5, Action 3)	Q(State 5, Action 4)
Target 6	Q(State 6, Action 0)	Q(State 6, Action 1)	Q(State 6, Action 2)	Q(State 6, Action 3)	Q(State 6, Action 4)
Target 7	Q(State 7, Action 0)	Q(State 7, Action 1)	Q(State 7, Action 2)	Q(State 7, Action 3)	Q(State 7, Action 4)
Target 8	Q(State 8, Action 0)	Q(State 8, Action 1)	Q(State 8, Action 2)	Q(State 8, Action 3)	Q(State 8, Action 4)
Target 9	Q(State 9, Action 0)	Q(State 9, Action 1)	Q(State 9, Action 2)	Q(State 9, Action 3)	Q(State 9, Action 4)
Target 10	Q(State 10, Action 0)	Q(State 10, Action 1)	Q(State 10, Action 2)	Q(State 10, Action 3)	Q(State 10, Action 4)

Figure 8: Batch of targets (2/3)

Remember that for the action that is played, the formula of the target must be this one:

$$R\left(s_{t_B}, a_{t_B}\right) + \gamma \max_a \left(Q\left(s_{t_B+1}, a\right)\right)$$

What we do in the following lines of code is to put this formula into the column of each action that was played within the 10 selected transitions. In other words, we get this:

	Action 0	Action 1	Action 2	Action 3	Action 4
Target 1	Q(State 1, Action 0)	r1+γmaxQ(next_state1,a)	Q(State 1, Action 2)	Q(State 1, Action 3)	Q(State 1, Action 4)
Target 2	Q(State 2, Action 0)	Q(State 2, Action 1)	Q(State 2, Action 2)	r2+γmaxQ(next_state2,a)	Q(State 2, Action 4)
Target 3	r3+γmaxQ(next_state3,a)	Q(State 3, Action 1)	Q(State 3, Action 2)	Q(State 3, Action 3)	Q(State 3, Action 4)
Target 4	Q(State 4, Action 0)	Q(State 4, Action 1)	r4+γmaxQ(next_state4,a)	Q(State 4, Action 3)	Q(State 4, Action 4)
Target 5	Q(State 5, Action 0)	Q(State 5, Action 1)	Q(State 5, Action 2)	Q(State 5, Action 3)	r5+γmaxQ(next_state5,a)
Target 6	Q(State 6, Action 0)	r6+γmaxQ(next_state6,a)	Q(State 6, Action 2)	Q(State 6, Action 3)	Q(State 6, Action 4)
Target 7	r7+γmaxQ(next_state7,a)	Q(State 7, Action 1)	Q(State 7, Action 2)	Q(State 7, Action 3)	Q(State 7, Action 4)
Target 8	Q(State 8, Action 0)	Q(State 8, Action 1)	Q(State 8, Action 2)	r8+γmaxQ(next_state8,a)	Q(State 8, Action 4)
Target 9	Q(State 9, Action 0)	Q(State 9, Action 1)	Q(State 9, Action 2)	Q(State 9, Action 3)	r9+γmaxQ(next_state9,a)
Target 10	Q(State 10, Action 0)	r10+γmaxQ(next_state10,a)	Q(State 10, Action 2)	Q(State 10, Action 3)	Q(State 10, Action 4)

Figure 9: Batch of targets (3/3)

In that example, **Action 1** was performed in the first transition (**Target 1**), **Action 3** was performed in the second transition (**Target 2**), **Action 0** was performed in the third transition (**Target 3**), and so on. Let's populate this in the following lines of code.

Line 35: We first start getting the $\max_a \left(Q\left(s_{t_B+1}, a\right)\right)$ part of the formula of the target:

$$R\left(s_{t_B}, a_{t_B}\right) + \gamma \max_a \left(Q\left(s_{t_B+1}, a\right)\right)$$

Line 36: We check if `game_over = 1`, meaning that the server has gone outside the allowed range of server temperatures. Because if it has, there's actually no next state (because we basically reset the environment by putting the server's temperature back into the optimal range so we start from a new state); and therefore we shouldn't consider $\max_a \left(Q\left(s_{t_B+1}, a\right)\right)$.

Line 37: In that case, we only keep the $R\left(s_{t_B}, a_{t_B}\right)$ part of the target.

Line 38: However, if the game is not over (`game_over = 0`)...

Line 39: We keep the whole formula of the target, but of course only for the action that was performed, meaning a_{t_B} here:

$$R\left(s_{t_B}, a_{t_B}\right) + \gamma \max_a \left(Q\left(s_{t_B+1}, a\right)\right)$$

Hence, we get the following batch of targets, as you saw earlier:

	Action 0	Action 1	Action 2	Action 3	Action 4
Target 1	Q(State 1, Action 0)	r1+γmaxQ(next_state1,a)	Q(State 1, Action 2)	Q(State 1, Action 3)	Q(State 1, Action 4)
Target 2	Q(State 2, Action 0)	Q(State 2, Action 1)	Q(State 2, Action 2)	r2+γmaxQ(next_state2,a)	Q(State 2, Action 4)
Target 3	r3+γmaxQ(next_state3,a)	Q(State 3, Action 1)	Q(State 3, Action 2)	Q(State 3, Action 3)	Q(State 3, Action 4)
Target 4	Q(State 4, Action 0)	Q(State 4, Action 1)	r4+γmaxQ(next_state4,a)	Q(State 4, Action 3)	Q(State 4, Action 4)
Target 5	Q(State 5, Action 0)	Q(State 5, Action 1)	Q(State 5, Action 2)	Q(State 5, Action 3)	r5+γmaxQ(next_state5,a)
Target 6	Q(State 6, Action 0)	r6+γmaxQ(next_state6,a)	Q(State 6, Action 2)	Q(State 6, Action 3)	Q(State 6, Action 4)
Target 7	r7+γmaxQ(next_state7,a)	Q(State 7, Action 1)	Q(State 7, Action 2)	Q(State 7, Action 3)	Q(State 7, Action 4)
Target 8	Q(State 8, Action 0)	Q(State 8, Action 1)	Q(State 8, Action 2)	r8+γmaxQ(next_state8,a)	Q(State 8, Action 4)
Target 9	Q(State 9, Action 0)	Q(State 9, Action 1)	Q(State 9, Action 2)	Q(State 9, Action 3)	r9+γmaxQ(next_state9,a)
Target 10	Q(State 10, Action 0)	r10+γmaxQ(next_state10,a)	Q(State 10, Action 2)	Q(State 10, Action 3)	Q(State 10, Action 4)

Figure 10: Batch of targets (3/3)

Line 40: At last, we `return` the final batches of `inputs` and `targets`.

That was epic—you've successfully created an artificial brain. Now that you've done it, we're ready to start the training.

Step 4: Training the AI

Now that our AI has a fully functional brain, it's time to train it. That's exactly what we do in this fourth Python implementation. You actually have a choice of two files to use for this:

1. `training_noearlystopping.py`, which trains your AI on a full 1000 epochs of 5-months period.

2. `training_earlystopping.py`, which trains your AI on 1000 epochs as well, but which can stop the training early if the performance no longer improves over the iterations. This technique is called **early stopping**.

Both these implementations are long, but very simple. We start by setting all the parameters, then we build the environment by creating an object of the `Environment()` class, then we build the brain of the AI by creating an object of the `Brain()` class, then we build the deep Q-learning model by creating an object of the `DQN()` class, and finally we launch the training connecting all these objects together over 1000 epochs of 5-month periods.

You'll notice in the training loop that we also do some exploration when performing the actions, performing some random actions from time to time. In our case, this will be done 30% of the time, since we use an exploration parameter $\epsilon = 0.3$, and then we force the AI to perform a random action when we draw a random value between 0 and 1 that is below $\epsilon = 0.3$. The reason we do some exploration is because it improves the deep reinforcement learning process, as we discussed in *Chapter 9, Going Pro with Artificial Brains – Deep Q-Learning,* and the reason we don't use Softmax in this project is just to give you a look at how to implement a different exploration method.

Later, you'll be introduced to another little improvement in the training_ noearlystopping.py file, where we use an early stopping technique which stops the training early if there's no improvement in the performance.

Let's highlight the new steps which still belong to our general AI framework/ Blueprint:

- **Step 4-1**: Building the environment by creating an object of the Environment class.

- **Step 4-2**: Building the artificial brain by creating an object of the Brain class.

- **Step 4-3**: Building the DQN model by creating an object of the DQN class.

- **Step 4-4**: Selecting the training mode.

- **Step 4-5**: Starting the training with a for loop over 100 epochs of 5-month periods.

- **Step 4-6**: During each epoch we repeat the whole deep Q-learning process, while also doing some exploration 30% of the time.

No early stopping

Ready to implement this? Maybe get a good coffee or tea first because this is going to be a bit long (88 lines of code, but easy ones!). We'll start without early stopping and then at the end I'll explain how to add the early stopping technique. The file to follow along with is training_noearlystopping.py. Since this is pretty long, let's do it section by section this time, starting with the first one:

```
1       # AI for Business - Minimize cost with Deep Q-Learning
2       # Training the AI without Early Stopping
3
4       # Importing the libraries and the other python files
5       import os
6       import numpy as np
7       import random as rn
```

```
8        import environment
9        import brain_nodropout
10       import dqn
```

Line 5: We import the `os` library, which will be used to set a seed for reproducibility so that if you run the training several times, you'll get the same result each time. You can, of course, choose to remove this when you tinker with the code yourself!

Line 6: We import the `numpy` library, since we'll work with `numpy` arrays.

Line 7: We import the `random` library, which we'll use to do some exploration.

Line 8: We import the `environment.py` file, implemented in Step 1, which contains the whole defined environment.

Line 9: We import the `brain_nodropout.py` file, our artificial brain without dropout that we implemented in Step 2. This contains the whole neural network of our AI.

Line 10: We import the `dqn.py` file implemented in Step 3, which contains the main parts of the deep Q-learning algorithm, including experience replay.

Moving on to the next section:

```
12       # Setting seeds for reproducibility
13       os.environ['PYTHONHASHSEED'] = '0'
14       np.random.seed(42)
15       rn.seed(12345)
16

17       # SETTING THE PARAMETERS
18       epsilon = .3
19       number_actions = 5
20       direction_boundary = (number_actions - 1) / 2
21       number_epochs = 100
22       max_memory = 3000
23       batch_size = 512
24       temperature_step = 1.5
25

26       # BUILDING THE ENVIRONMENT BY SIMPLY CREATING AN OBJECT OF THE
         ENVIRONMENT CLASS
27       env = environment.Environment(optimal_temperature = (18.0,
         24.0), initial_month = 0, initial_number_users = 20, initial_
         rate_data = 30)
```

```
28

29    # BUILDING THE BRAIN BY SIMPLY CREATING AN OBJECT OF THE BRAIN
      CLASS
30    brain = brain_nodropout.Brain(learning_rate = 0.00001, number_
      actions = number_actions)
31

32    # BUILDING THE DQN MODEL BY SIMPLY CREATING AN OBJECT OF THE DQN
      CLASS
33    dqn = dqn.DQN(max_memory = max_memory, discount = 0.9)
34

35    # CHOOSING THE MODE
36    train = True
```

Lines 13, 14, and 15: We set seeds for reproducibility, to get the same results after several rounds of training. This is really only important so you can reproduce your findings—if you don't need to do that, some people prefer them and others don't. If you don't want the seeds you can just remove them.

Line 18: We introduce the exploration factor $\in$, and we set it to 0.3, meaning that there will be 30% of exploration (performing random actions) vs. 70% of exploitation (performing the actions of the AI).

Line 19: We set the number of actions to 5.

Line 20: We set the direction boundary, meaning the action index below which we cool down the server, and above which we heat up the server. Since actions 0 and 1 cool down the server, and actions 3 and 4 heat up the server, that direction boundary is (5-1)/2 = 2, which corresponds to the action that transfers no heat to the server (action 2).

Line 21: We set the number of training epochs to 100.

Line 22: We set the memory capacity, meaning its maximum size, to 3000.

Line 23: We set the batch size to 512.

Line 24: We introduce the temperature step, meaning the absolute temperature change that the AI cause onto the server by playing actions 0, 1, 3, or 4. And that's of course 1.5°C.

Line 27: We create the `environment` object, as an instance of the `Environment` class which we call from the `environment` file. Inside this `Environment` class, we enter all the arguments of the `init` method:

```
optimal_temperature = (18.0, 24.0),
initial_month = 0,
initial_number_users = 20,
initial_rate_data = 30
```

Line 30: We create the `brain` object as an instance of the `Brain` class, which we call from the `brain_nodropout` file. Inside this `Brain` class, we enter all the arguments of the `init` method:

```
learning_rate = 0.00001,
number_actions = number_actions
```

Line 33: We create the `dqn` object as an instance of the `DQN` class, which we call from the `dqn` file. Inside this `DQN` class we enter all the arguments of the `init` method:

```
max_memory = max_memory,
discount = 0.9
```

Line 36: We set the training mode to `True`, because the next code section will contain the big `for` loop that performs all the training.

All good so far? Don't forget to take a break or a step back by reading the previous paragraphs again anytime you feel a bit overwhelmed or lost.

Now let's begin the big training loop; that's the last code section of this file:

```
38      # TRAINING THE AI
39      env.train = train
40      model = brain.model
41      if (env.train):
42          # STARTING THE LOOP OVER ALL THE EPOCHS (1 Epoch = 5 Months)
43          for epoch in range(1, number_epochs):
44              # INITIALIAZING ALL THE VARIABLES OF BOTH THE
        ENVIRONMENT AND THE TRAINING LOOP
45              total_reward = 0
46              loss = 0.
47              new_month = np.random.randint(0, 12)
48              env.reset(new_month = new_month)
49              game_over = False
50              current_state, _, _ = env.observe()
51              timestep = 0
52              # STARTING THE LOOP OVER ALL THE TIMESTEPS (1 Timestep =
        1 Minute) IN ONE EPOCH
```

```
53              while ((not game_over) and timestep <= 5 * 30 * 24 *
        60):
54                  # PLAYING THE NEXT ACTION BY EXPLORATION
55                  if np.random.rand() <= epsilon:
56                      action = np.random.randint(0, number_actions)
57                      if (action - direction_boundary < 0):
58                          direction = -1
59                      else:
60                          direction = 1
61                      energy_ai = abs(action - direction_boundary) *
        temperature_step
62                  # PLAYING THE NEXT ACTION BY INFERENCE
63                  else:
64                      q_values = model.predict(current_state)
65                      action = np.argmax(q_values[0])
66                      if (action - direction_boundary < 0):
67                          direction = -1
68                      else:
69                          direction = 1
70                      energy_ai = abs(action - direction_boundary) *
        temperature_step
71                  # UPDATING THE ENVIRONMENT AND REACHING THE NEXT
        STATE
72                  next_state, reward, game_over = env.update_
        env(direction, energy_ai, ( new_month + int(timestep/(30*24*60))
        ) % 12)
73                  total_reward += reward
74                  # STORING THIS NEW TRANSITION INTO THE MEMORY
75                  dqn.remember([current_state, action, reward, next_
        state], game_over)
76                  # GATHERING IN TWO SEPARATE BATCHES THE INPUTS AND
        THE TARGETS
77                  inputs, targets = dqn.get_batch(model, batch_size =
        batch_size)
78                  # COMPUTING THE LOSS OVER THE TWO WHOLE BATCHES OF
        INPUTS AND TARGETS
79                  loss += model.train_on_batch(inputs, targets)
80                  timestep += 1
81                  current_state = next_state
82              # PRINTING THE TRAINING RESULTS FOR EACH EPOCH
83              print("\n")
```

```
84              print("Epoch: {:03d}/{:03d}".format(epoch, number_
        epochs))
85                  print("Total Energy spent with an AI: {:.0f}".
        format(env.total_energy_ai))
86                  print("Total Energy spent with no AI: {:.0f}".
        format(env.total_energy_noai))
87              # SAVING THE MODEL
88              model.save("model.h5")
```

Line 39: We set the `env.train` object variable (this is a variable of our `environment` object) to the value of the `train` variable entered just before, which is of course equal to `True`, meaning we are indeed in training mode.

Line 40: We get the model from our `brain` object. This model contains the whole architecture of the neural network, plus its optimizer. It also has extra practical tools, like for example the `save` and `load` methods, which will allow us respectively to save the weights after the training or load them anytime in the future.

Line 41: If we are in training mode…

Line 43: We start the main training `for` loop, iterating the training epochs from 1 to 100.

Line 45: We set the total reward (total reward accumulated over the training iterations) to 0.

Line 46: We set the loss to 0 (0 because the loss will be a `float`).

Line 47: We set the starting month of the training, called new_month, to a random integer between 0 and 11. For example, if the random integer is 2, we start the training in March.

Line 48: By calling the `reset` method from our `env` object of the `Environment` class built in Step 1, we reset the environment starting from that new_month.

Line 49: We set the game_over variable to `False`, because we're starting in the allowed range of server temperatures.

Line 50: By calling the `observe` method from our `env` object of the `Environment` class built in Step 1, we get the current state only, which is our starting state.

Line 51: We set the first `timestep` to 0. This is the first minute of the training.

Line 53: We start the `while` loop that will iterate all the timesteps (minutes) for the whole period of the epoch, which is 5 months. Therefore, we iterate through 5 * 30 * 24 * 60 minutes; that is, 216,000 timesteps.

If, however, during those timesteps we go outside the allowed range of server temperatures (that is, if game_over = 1), then we stop the epoch and we start a new one.

Lines 55 to 61 make sure the AI performs a random action 30% of the time. This is exploration. The trick to it in this case is to sample a random number between 0 and 1, and if this random number is between 0 and 0.3, the AI performs a random action. That means the AI will perform a random action 30% of the time, because this sampled number has a 30% chance to be between 0 and 0.3.

Line 55: If a sampled number between 0 and 1 is below $\epsilon = 0.3$...

Line 56: ... we play a random action index from 0 to 4.

Line 57: Now that we've just performed an action, we compute the direction and the energy spent; remember that they're are the required arguments of the update_env method of the Environment class, which we'll call later to update the environment. The AI distinguishes between two cases by checking if the action is below or above the direction boundary of 2. If the action is below the direction boundary of 2, meaning the AI cools down the server...

Line 58: ...then the heating direction is equal to -1 (cooling down).

Line 59 and 60: Else the heating direction is equal to +1 (heating up).

Line 61: We compute the energy spent by the AI onto the server, which according to Assumption 2 is:

$$| action - direction\_boundary | * temperature\_step = | action - 2 | * 1.5\ Joules$$

For example, if the action is 4, then the AI heats up the server by 3°C, and so according to Assumption 2 the energy spent is 3 Joules. And we check indeed that $|4-2|*1.5 = 3$.

Line 63: Now we play the actions by inference, meaning directly from our AI's predictions. The inference starts from the else statement, which corresponds to the if statement of line 55. This else corresponds to the situation where the sampled number is between 0.3 and 1, which happens 70% of the time.

Line 64: By calling the predict method from our model object (predict is a pre-built method of the Model class), we get the five predicted Q-values from our AI model.

Line 65: Using the argmax function from numpy, we select the action that has the maximum Q-value among the five predicted ones at Line 64.

Lines 66 to 70: We do exactly the same as in Lines 57 to 61, but this time with the action performed by inference.

Line 72: Now we have everything ready to update the environment. We call the big `update_env` method made in the `Environment` class of Step 1, by inputting the heating direction, the energy spent by the AI, and the month we're in at that specific timestep of the `while` loop. We get in return the next state, the reward received, and whether the game is over (that is, whether or not we went outside the optimal range of server temperatures).

Line 73: We add this last reward received to the total reward.

Line 75: By calling the `remember` method from our dqn object of the DQN class built in Step 3, we store the new transition [[`current_state`, `action`, `reward`, `next_state`], `game_over`] into the memory.

Line 77: By calling the `get_batch` method from our dqn object of the DQN class built in Step 3, we create two separate batches of `inputs` and `targets`, each one having 512 elements (since `batch_size` = 512).

Line 79: By calling the `train_on_batch` method from our model object (`train_on_batch` is a pre-built method of the `Model` class), we compute the loss error between the predictions and the targets over the whole batch. As a reminder, this loss error is the mean-squared error loss. Then in this same line, we add this loss error to the total loss of the epoch, in case we want to check how this total loss evolves over the epochs during the training.

Line 80: We increment the `timestep`.

Line 81: We update the current state, which becomes the new state reached.

Line 83: We print a new line to separate out the training results so we can look them over easily.

Line 84: We print the epoch reached (the one we are in at this specific moment of the main training `for` loop).

Line 85: We print the total energy spent by the AI over that specific epoch (the one we are in at this specific moment of the main training `for` loop).

Line 86: We print the total energy spent by the server's integrated cooling system over that same specific epoch.

Line 88: We save the model's weights at the end of the training, in order to load them in the future, anytime we want to use our pre-trained model to regulate a server's temperature.

That's it for training our AI without early stopping; now let's have a look at what you'd need to change to implement it.

Early stopping

Now open the `training_earlystopping.py` file. Compare it to the previous file; all the lines of code from 1 to 40 are the same. Then, in the last code section, `TRAINING THE AI`, we have the same process, to which is added the early stopping technique. As a reminder, it consists of stopping the training if there's no more improvement of the performance, which could be assessed two different ways:

1. If the total reward of an epoch no longer increases much over the epochs.
2. If the total loss of an epoch no longer decreases much over the epochs.

Let's see how we do this.

First, we introduce four new variables just before the main training `for` loop:

```
38      # TRAINING THE AI
39      env.train = train
40      model = brain.model
41      early_stopping = True
42      patience = 10
43      best_total_reward = -np.inf
44      patience_count = 0
45      if (env.train):
46          # STARTING THE LOOP OVER ALL THE EPOCHS (1 Epoch = 5 Months)
47          for epoch in range(1, number_epochs):
```

Line 41: We introduce a new variable, `early_stopping`, which is set equal to `True` if we decide to activate the early stopping technique, meaning if we decide to stop the training when the performance no longer improves.

Line 42: We introduce a new variable, `patience`, which is the number of epochs we wait without performance improvement before stopping the training. Here we choose a patience of `10` epochs, which means that if the best total reward of an epoch doesn't increase during the next 10 epochs, we will stop the training.

Line 43: We introduce a new variable, `best_total_reward`, which is the best total reward recorded over a full epoch. If we don't beat that best total reward before 10 epochs go by, the training stops. It's initialized to `-np.inf`, which represents `-infinity`. That's just a trick to say that nothing can be lower than that best total reward at the beginning. Then as soon as we get the first total reward over the first epoch, `best_total_reward` becomes that first total reward.

Line 44: We introduce a new variable, `patience_count`, which is a counter starting from `0`, and is incremented by 1 each time the total reward of an epoch doesn't beat the best total reward. If `patience_count` reaches 10 (the patience), we stop the training. And if one epoch beats the best total reward, `patience_count` is reset to 0.

Then, the main training `for` loop is the same as before, but just before saving the model we add the following:

```
91              # EARLY STOPPING
92              if (early_stopping):
93                  if (total_reward <= best_total_reward):
94                      patience_count += 1
95                  elif (total_reward > best_total_reward):
96                      best_total_reward = total_reward
97                      patience_count = 0
98                  if (patience_count >= patience):
99                      print("Early Stopping")
100                     break
101             # SAVING THE MODEL
102             model.save("model.h5")
```

Line 92: If the `early_stopping` variable is `True`, meaning if the early stopping technique is activated...

Line 93: And if the total reward of the current epoch (we are still in the main training `for` loop that iterates the epochs) is lower than the best total reward of an epoch obtained so far...

Line 94: ...we increment the `patience_count` variable by `1`.

Line 95: However, if the total reward of the current epoch is higher than the best total reward of an epoch obtained so far...

Line 96: ...we update the best total reward, which becomes that new total reward of the current epoch.

Line 97: ...and we reset the `patience_count` variable to `0`.

Line 98: Then in a new `if` condition, we check that if the `patience_count` variable goes higher than the patience of 10...

Line 99: ...we print `Early Stopping`,

Line 100: ...and we stop the main training `for` loop with a `break` statement.

That's the whole thing. Easy and intuitive, right? Now you know how to implement early stopping.

After executing the code (I'll explain how to run this in a bit), we'll already see some good performances from our AI during the training, spending less energy than the server's integrated cooling system most of the time. But that's only training; now we need to see if we get good performance from the AI on a new 1-year simulation. That's where our next and final Python file comes into play.

Step 5 – Testing the AI

Now we need to test the performance of our AI in a brand-new situation. To do so, we run a 1-year simulation in inference mode, meaning that there's no training happening at any time. Our AI only returns predictions over a full year of simulation. Then, thanks to our environment object, in the end we'll be able to see the total energy spent by the AI over the full year, as well as the total energy that would have been spent in the exact same year by the server's integrated cooling system. Finally, we compare these two total energies spent, by computing their relative difference (in %) which shows us precisely the total energy saved by the AI. Buckle up for the final results—we'll reveal them very soon!

In terms of the AI blueprint, for the testing implementation we have almost the same process as the training implementation, except that this time we don't need to create a `brain` object nor a `DQN` model object; and, of course, we won't run the deep Q-learning process over some training epochs. However, we do have to create a new `environment` object, and instead of creating a `brain`, we'll load our artificial brain with its pre-trained weights from the previous training that we executed in Step 4 – Training the AI. Let's take a look at the final sub-steps of this final part of the AI framework/Blueprint:

- **Step 5-1**: Build a new environment by creating an object of the `Environment` class.

- **Step 5-2**: Load the artificial brain with its pre-trained weights from the previous training.

- **Step 5-3**: Choose the inference mode.

- **Step 5-4**: Start the 1-year simulation.

- **Step 5-5**: In each iteration (each minute), our AI only performs the action that results from its prediction, and no exploration or deep Q-learning training happens whatsoever.

The implementation is a piece of cake to understand. It's actually the same as the training file, except that:

1. Instead of creating a `brain` object from the `Brain` class, we load the pre-trained weights resulting from the training.

2. Instead of running a training loop over 100 epochs of 5-month periods, we run an inference loop over a single 12-month period. Inside this inference loop, you'll find exactly the same code as the inference part of the training `for` loop. You've got this!

Have a look at the full testing implementation in the following code:

```
# AI for Business - Minimize cost with Deep Q-Learning
# Testing the AI

# Installing Keras
# conda install -c conda-forge keras

# Importing the libraries and the other python files
import os
import numpy as np
import random as rn
from keras.models import load_model
import environment

# Setting seeds for reproducibility
os.environ['PYTHONHASHSEED'] = '0'
np.random.seed(42)
rn.seed(12345)

# SETTING THE PARAMETERS
number_actions = 5
direction_boundary = (number_actions - 1) / 2
temperature_step = 1.5

# BUILDING THE ENVIRONMENT BY SIMPLY CREATING AN OBJECT OF THE
ENVIRONMENT CLASS
env = environment.Environment(optimal_temperature = (18.0, 24.0),
initial_month = 0, initial_number_users = 20, initial_rate_data = 30)

# LOADING A PRE-TRAINED BRAIN
model = load_model("model.h5")

# CHOOSING THE MODE
train = False
```

```
# RUNNING A 1 YEAR SIMULATION IN INFERENCE MODE
env.train = train
current_state, _, _ = env.observe()
for timestep in range(0, 12 * 30 * 24 * 60):
    q_values = model.predict(current_state)
    action = np.argmax(q_values[0])
    if (action - direction_boundary < 0):
        direction = -1
    else:
        direction = 1
    energy_ai = abs(action - direction_boundary) * temperature_step
    next_state, reward, game_over = env.update_env(direction, energy_
ai, int(timestep / (30 * 24 * 60)))
    current_state = next_state

# PRINTING THE TRAINING RESULTS FOR EACH EPOCH
print("\n")
print("Total Energy spent with an AI: {:.0f}".format(env.total_energy_
ai))
print("Total Energy spent with no AI: {:.0f}".format(env.total_energy_
noai))
print("ENERGY SAVED: {:.0f} %".format((env.total_energy_noai - env.
total_energy_ai) / env.total_energy_noai * 100))
```

Everything's more or less the same as before; we just removed the parts related to the training.

The demo

Given the different files we have, make sure to understand that there are four possible ways to run the program:

1. Without dropout and without early stopping
2. Without dropout and with early stopping
3. With dropout and without early stopping
4. With dropout and with early stopping

Then, for each of these four combinations, the way to run this is the same: we first execute the training file, and then the testing file. In this demo section, we'll execute the 4th option, with both dropout and early stopping.

Now how do we run this? We have two options: with or without Google Colab.

I'll explain how to do it on Google Colab, and I'll even give you a Google Colab file where you only have to hit the play button. For those of you who want to execute this without Colab, on your favorite Python IDE, or through the terminal, let me explain how it's done. It's easy; you just need to download the main repository from GitHub, then in your Python IDE set the right working directory folder, which is the `Chapter 11` folder, and then run the following two files in this order:

1. `training_earlystopping.py`, inside which you should make sure to import `brain_dropout` at line 9. This will execute the training, and you'll have to wait until that finishes (which will take about 10 minutes).

2. `testing.py`, which will test the model on one full year of data.

Now, back to Google Colab. First, open a new Colaboratory file, and call it **Deep Q-Learning for Business**. Then add all your files from the `Chapter 11` folder of GitHub into this Colaboratory file, right here:

Figure 11: Google Colab – Step 1

Unfortunately, it's not easy to add the different files manually. You can only do this by using the os library, which we won't bother with. Instead, copy-paste the five Python implementations in five different cells of our Colaboratory file, in the following order:

1. A first cell containing the whole `environment.py` implementation.

2. A second cell containing the whole `brain_dropout.py` implementation.

3. A third cell containing the whole `dqn.py` implementation.

4. A fourth cell containing the whole `training_earlystopping.py` implementation.

5. And a last cell containing the whole `testing.py` implementation.

Here's what it looks like, after adding some snazzy titles:

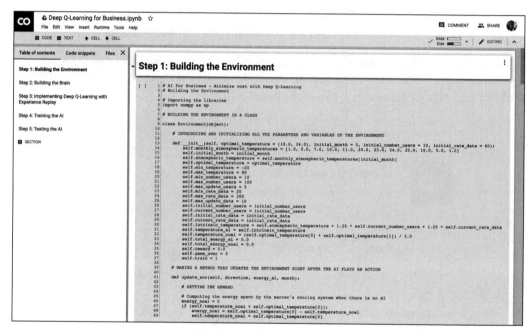

Figure 12: Google Colab – Step 2

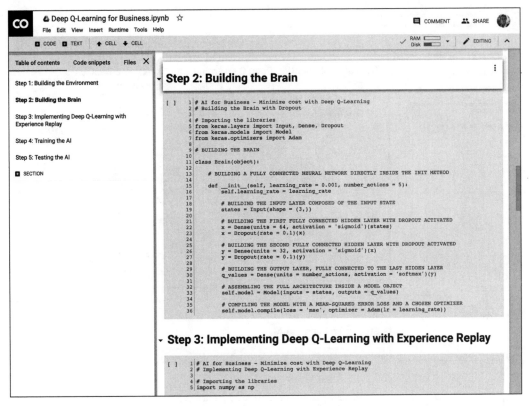

Figure 13: Google Colab – Step 3

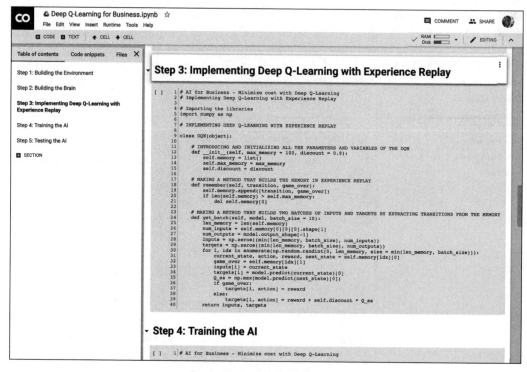

Figure 14: Google Colab – Step 4

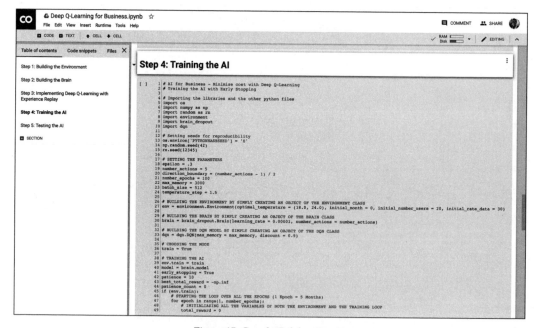

Figure 15: Google Colab – Step 5

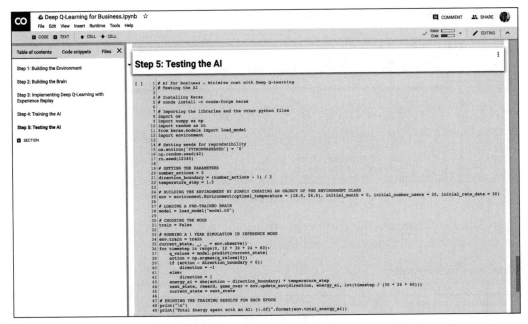

Figure 16: Google Colab – Step 6

Now before we execute each of these cells in the order one through five, we need to remove the `import` commands of the Python files. The reason for this is that now that the implementations are in cells, they're like a single Python implementation, and we don't have to import the interdependent files in every single cell. First, remove the following three different rows in the training file:

Step 4: Training the AI

```
1  # AI for Business - Minimize cost with Deep Q-Learning
2  # Training the AI with Early Stopping
3
4  # Importing the libraries and the other python files
5  import os
6  import numpy as np
7  import random as rn
8  import environment
9  import brain_dropout
10 import dqn
11
```

Figure 17: Google Colab – Step 7

After doing that, we end up with this:

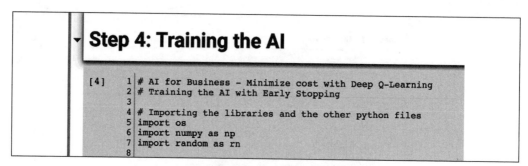

```
Step 4: Training the AI

[4]   1 # AI for Business - Minimize cost with Deep Q-Learning
      2 # Training the AI with Early Stopping
      3
      4 # Importing the libraries and the other python files
      5 import os
      6 import numpy as np
      7 import random as rn
      8
```

Figure 18: Google Colab – Step 8

Then, since we removed these imports, we also have to remove the three filenames for the `environment`, the `brain`, and the `dqn`, when creating the objects:

First the environment:

```
21 temperature_step = 1.5
22
23 # BUILDING THE ENVIRONMENT BY SIMPLY CREATING AN OBJECT OF THE ENVIRONMENT CLASS
24 env = environment.Environment(optimal_temperature = (18.0, 24.0), initial_month = 0, in
25
```

Figure 19: Google Colab – Step 9

Then the brain:

```
24 env = environment.Environment(optimal_temperature = (18.0, 24.0), initial_month = 0,
25
26 # BUILDING THE BRAIN BY SIMPLY CREATING AN OBJECT OF THE BRAIN CLASS
27 brain = brain_dropout.Brain(learning_rate = 0.00001, number_actions = number_actions)
28
29 # BUILDING THE DQN MODEL BY SIMPLY CREATING AN OBJECT OF THE DQN CLASS
30 dqn = dqn.DQN(max_memory = max_memory, discount = 0.9)
31
```

Figure 20: Google Colab – Step 10

And finally the dqn:

```
28
29 # BUILDING THE DQN MODEL BY SIMPLY CREATING AN OBJECT OF THE DQN CLASS
30 dqn = dqn.DQN(max_memory = max_memory, discount = 0.9)
31
32 # CHOOSING THE MODE
```

Figure 21: Google Colab – Step 11

Now the training file's good to go. In the testing file, we just have to remove two things, the environment import at line 12:

```
 8 import os
 9 import numpy as np
10 import random as rn
11 from keras.models import load_model
12 import environment
13
```

Figure 22: Google Colab – Step 12

and the environment. at row 25:

```
24 # BUILDING THE ENVIRONMENT BY SIMPLY CREATING AN OBJECT OF THE ENVIRONMENT CLASS
25 env = environment.Environment(optimal_temperature = (18.0, 24.0), initial_month = 0,
26
27 # LOADING A PRE-TRAINED BRAIN
28 model = load_model("model.h5")
```

Figure 23: Google Colab – Step 13

That's it; now you're all set! You're ready to literally hit the play button on each of the cells from top to the bottom.

First, execute the first cell. After executing it, no output is displayed. That's fine!

Then execute the second cell:

Using TensorFlow backend.

After executing it, you can see the output Using TensorFlow backend.

Then execute the third cell, after which no output is displayed.

Now it gets a bit exciting! You're about to execute the training, and follow the training performance in real time. Do this by executing the fourth cell. After executing it, the training launches, and you should see the following results:

```
...  WARNING:tensorflow:From /usr/local/lib/python3.6/dist-packages/tensorflow/python/framework/
     Instructions for updating:
     Colocations handled automatically by placer.
     WARNING:tensorflow:From /usr/local/lib/python3.6/dist-packages/keras/backend/tensorflow_bac
     Instructions for updating:
     Please use `rate` instead of `keep_prob`. Rate should be set to `rate = 1 - keep_prob`.
     WARNING:tensorflow:From /usr/local/lib/python3.6/dist-packages/tensorflow/python/ops/math_o
     Instructions for updating:
     Use tf.cast instead.

     Epoch: 001/100
     Total Energy spent with an AI: 30
     Total Energy spent with no AI: 146

     Epoch: 002/100
     Total Energy spent with an AI: 0
     Total Energy spent with no AI: 0

     Epoch: 003/100
     Total Energy spent with an AI: 4
     Total Energy spent with no AI: 22

     Epoch: 004/100
     Total Energy spent with an AI: 28
     Total Energy spent with no AI: 116

     Epoch: 005/100
     Total Energy spent with an AI: 46
     Total Energy spent with no AI: 224
```

Figure 24: The output

Don't worry about those warnings, everything's running the way it should. Since early stopping is activated, you'll reach the end of the training way before the 100 epochs, at the 15th epoch:

```
     Epoch: 012/100
     Total Energy spent with an AI: 0
     Total Energy spent with no AI: 2

     Epoch: 013/100
     Total Energy spent with an AI: 3
     Total Energy spent with no AI: 0

     Epoch: 014/100
     Total Energy spent with an AI: 0
     Total Energy spent with no AI: 0

     Epoch: 015/100
     Total Energy spent with an AI: 0
     Total Energy spent with no AI: 0
     Early Stopping
```

Figure 25: The output at the 15th epoch

Note that the pre-trained weights are saved in **Files**, in the `model.h5` file:

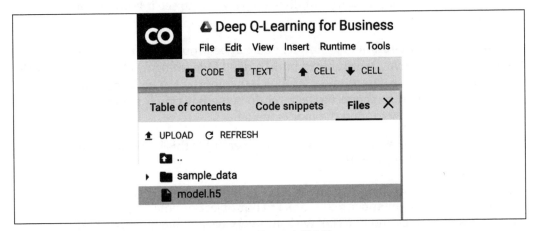

Figure 26: The model.h5 file

The training results look promising. Most of the time the AI spends less energy than the alternative server's integrated cooling system. Check that this is still the case with a full test, on one new year of simulation.

Execute the final cell and when it finishes running, (which takes approximately 3 minutes), you obtain in the printed results that the total energy consumption saved by the AI is...

```
Total Energy spent with an AI: 261985
Total Energy spent with no AI: 1978293
ENERGY SAVED: 87%
```

Total Energy saved by the AI = 87%

That's a lot of energy saved! Google DeepMind achieved similarly impressive results in 2016. If you look up the results by searching "DeepMind reduces Google cooling bill," you'll see that the result they achieved was 40%. Not bad! Of course, let's be critical: their server/ data center environment is much more complex than our server environment and has many more parameters, so even though they have one of the most talented AI teams in the world, they could only reduce the cooling bill by less than 50%.

Our environment's very simple, and if you dig into it (which I recommend you do) you'll likely find that the variations of users and data, and therefore the variation of temperature, follow a uniform distribution. Accordingly, the server's temperature usually stays around the optimal range of temperatures. The AI understands that well, and thus chooses most of the time to take no action and cause no change of temperature, thus consuming very little energy.

I highly recommend that you play around with your server cooling model; make it as complex as you like, and try out different rewards to see if you can cause different behaviors.

Even though our environment is simple, you can be proud of your achievement. What matters is that you were able to build a deep Q-learning model for a real-world business problem. The environment itself is less important; what's most important is that you know how to connect a deep reinforcement learning model to an environment, and how to train the model inside.

Now, after your successes with the self-driving car plus this business application, you know how to do just that!

What we've built is excellent for our business client, as our AI will seriously reduce their costs. Remember that thanks to our object-oriented structure (working with classes and objects), we could very easily take the objects created in this implementation for one server, and then plug them into other servers, so that in the end we end up lowering the total energy consumption of a whole data center! That's how Google saved billions of dollars in energy-related costs, thanks to the DQN model built by their DeepMind AI.

My heartiest congratulations to you for smashing this new application. You've just made huge progress with your AI skills.

Finally, here's the link to the Colaboratory file with this whole implementation as promised. You don't have to install anything, Keras and NumPy are already pre-installed (this is the beauty of Google Colab!):

```
https://colab.research.google.com/drive/1KGAoT7S60OC3UGHNnrr_
FuN5Hcil0cHk
```

Before we finish this chapter and move onto the world of deep convolutional Q-learning, let me give you a useful recap of the whole general AI blueprint when building a deep reinforcement learning model.

Recap – The general AI framework/ Blueprint

Let's recap the whole AI Blueprint, so that you can print it out and put it on your wall.

Step 1: Building the environment

- **Step 1-1**: Introducing and initializing all the parameters and variables of the environment.

- **Step 1-2**: Making a method that updates the environment right after the AI plays an action.
- **Step 1-3**: Making a method that resets the environment.
- **Step 1-4**: Making a method that gives us at any time the current state, the last reward obtained, and whether the game is over.

Step 2: Building the brain

- **Step 2-1**: Building the input layer composed of the input states.
- **Step 2-2**: Building the hidden layers with a chosen number of these layers and neurons inside each, fully connected to the input layer and between each other.
- **Step 2-3**: Building the output layer, fully connected to the last hidden layer.
- **Step 2-4**: Assembling the full architecture inside a model object.
- **Step 2-5**: Compiling the model with a mean squared error loss function and a chosen optimizer (a good one is Adam).

Step 3: Implementing the deep reinforcement learning algorithm

- **Step 3-1**: Introducing and initializing all the parameters and variables of the DQN model.
- **Step 3-2**: Making a method that builds the memory in experience replay.
- **Step 3-3**: Making a method that builds and returns two batches of 10 inputs and 10 targets.

Step 4: Training the AI

- **Step 4-1**: Building the environment by creating an object of the Environment class built in Step 1.
- **Step 4-2**: Building the artificial brain by creating an object of the Brain class built in Step 2.
- **Step 4-3**: Building the DQN model by creating an object of the DQN class built in Step 3.
- **Step 4-4**: Choosing the training mode.
- **Step 4-5**: Starting the training with a for loop over a chosen number of epochs.
- **Step 4-6**: During each epoch we repeat the whole deep Q-learning process, while also doing some exploration 30% of the time.

Step 5: Testing the AI

- **Step 5-1**: Building a new environment by creating an object of the `Environment` class built in Step 1.

- **Step 5-2**: Loading the artificial brain with its pre-trained weights from the previous training.

- **Step 5-3**: Choosing the inference mode.

- **Step 5-4**: Starting the simulation.

- **Step 5-5**: At each iteration (each minute), our AI only plays the action that results from its prediction, and no exploration or deep Q-learning training is happening whatsoever.

Summary

In this chapter you re-applied deep Q-learning to a new business problem. You were supposed to find the best strategy to cool down and heat up the server. Before you started defining the AI strategy, you had to make some assumptions about your environment, for example the way the temperature is calculated. As inputs to your ANN, you had information about the server at any given time, like the temperature and data transmission. As outputs, your AI predicted whether to cool down or heat up our server by a certain amount. The reward was the energy saved with respect to the other, traditional cooling system. Your AI was able to save 87% energy.

12

Deep Convolutional Q-Learning

Now that you understand how **Artificial Neural Networks** (**ANNs**) work, you're ready to tackle an incredibly useful tool, mostly used when dealing with images—**Convolutional Neural Networks** (**CNNs**). To put it simply, CNNs allow your AI to see images in real time as if it had eyes.

We will tackle them in the following steps:

1. What are CNNs used for?
2. How do CNNs work?
3. Convolution
4. Max pooling
5. Flattening
6. Full connection

Once you've understood those steps, you'll understand CNNs, and how they can be used in deep convolutional Q-learning.

What are CNNs used for?

CNNs are mostly used with images or videos, and sometimes with text to tackle **Natural Language Processing** (**NLP**) problems. They are often used in object recognition, for example, predicting whether there is a cat or a dog in a picture or video. They are also often used with deep Q-learning (which we will discuss later on), when the environment returns 2D states of itself, for example, when we are trying to build a self-driving car that reads outputs from cameras around it.

Remember the example in *Chapter 9, Going Pro with Artificial Brains - Deep Q-Learning,* where we were predicting houses' prices. As inputs, we had all of the values that define a house (area, age, number of bedrooms, and so on), and as output, we had the price of a house. In the case of CNNs, things are very similar. For example, if we wanted to solve the same problem using CNNs, we would have images of houses as inputs and the price of a house as output.

This diagram should illustrate what I mean:

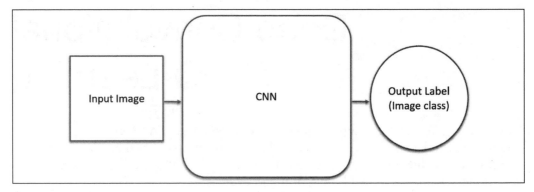

Figure 1: Input Image – CNN – Output Label

As you can see, the input is an image that flows through a CNN and comes out as an output. In the case of this diagram, the output is a class to which the image corresponds. What do I mean by a class? For example, if we wanted to predict whether the inputted image is a smiling face or a sad face, then one class would be *smiling face*, and the other would be *sad face*. Our output should then correctly decide to which class the input image corresponds.

Speaking of happy and sad faces, here's a diagram that represents it in more detail:

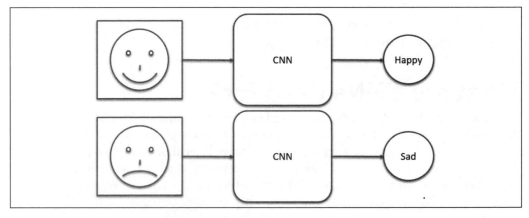

Figure 2: Two different classes to predict (Happy or Sad)

In the preceding example, we've run two images through a CNN. The first one is a smiling face and the other one is a sad face. As I mentioned before, our network predicts whether the image is a happy or a sad face.

I can imagine what you're thinking right now: how does it all work? What's inside this black box we call a CNN? I'll answer these questions in the following sections.

How do CNNs work?

Before we can go deep into the structure of CNNs, we need to understand a couple of points. I will introduce you to the first point with a question: how many dimensions does a colored RGB image have?

The answer may surprise you: it's 3!

Why? Because every RGB image is, in fact, represented by three 2D images, each one corresponding to a color in RGB architecture. So, there is one image corresponding to red, one corresponding to green, and one to blue. Grayscale images are only 2D, because they are represented by only one scale as there are no colors. The following diagram should make it clearer:

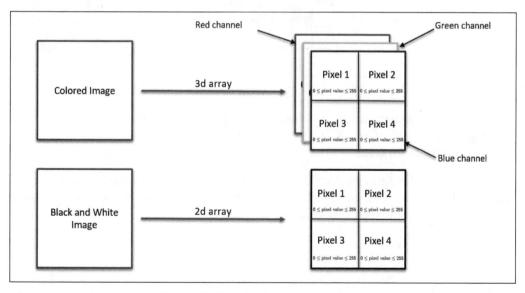

Figure 3: RGB versus black and white images

As you can see, a colored image is represented by a 3D array. Each color has its own layer in the picture, and this layer is called a **channel**. A grayscale (black and white) image only has one channel and is, therefore, a 2D array.

As you probably know, images are made out of pixels. Each of these is represented by a value that ranges from 0 to 255, where 0 is a pixel turned off and 255 is a fully bright pixel. It's important to understand that when we say that a pixel has the value (255, 255, 0), then that means this pixel is fully bright on the red and green channel and turned off on the blue channel.

From now on, to understand everything better, we'll be dealing with very simple images. In fact, our images will be grayscale (1 channel, 2D) and the pixels will either be fully bright or turned off. In order to make pictures easier to read, we'll assign 1 to a turned off pixel (black) and 0 to a fully bright one (white).

Going back to the case of happy and sad faces, this is what our 2D array representing a happy face would look like:

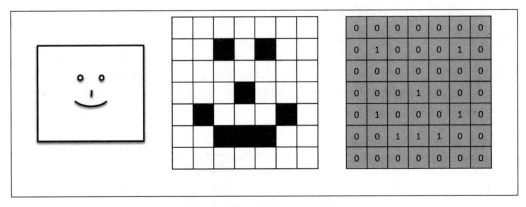

Figure 4: The pixel representation

As you can see, we have an array where **0** corresponds to a white pixel and **1** corresponds to a black pixel. The picture on the right is our smiling face represented by an array.

Now that we understand the foundations and that we've simplified the problem, we're ready to tackle CNNs. In order to fully understand them, we need to split our learning into the four steps that make up a CNN:

1. Convolution
2. Max pooling
3. Flattening
4. Full connection

Now we'll get to know each of these four steps one by one.

Step 1 – Convolution

This is the first crucial step of every CNN. In convolution, we apply something called **feature detectors** to the inputted image. Why do we have to do so? This is because all images contain certain features that define what is in the picture. For example, to recognize which face is sad and which one is happy, we need to understand the meaning of the shape of the mouth, which is a feature of this image. It's easier to understand this from a diagram:

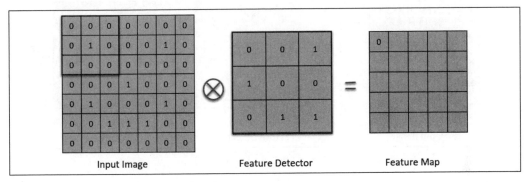

Figure 5: Step 1 – Convolution (1/5)

In the preceding diagram, we applied a feature detector, also known as a filter, to the smiling face we had as input. As you can see, a filter is a 2D array with some values inside. When we apply this feature detector to the image it covers (in this case it is a 3 x 3 grid), we check how many pixels from this part of the image match the filter's pixels. Then we put this number into a new 2D array called **feature map**. In other words, the more a part of the picture matches the picture detector, the higher the number we put into the feature map.

Next, we *slide* the feature detector across the entire image. In the next iteration, this is what will happen:

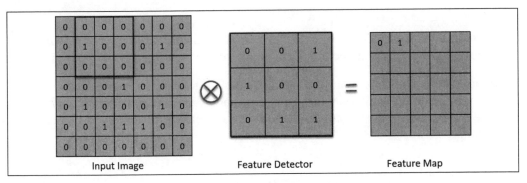

Figure 6: Step 1 – Convolution (2/5)

As you can see, we slide the filter one place to the right. This time, one pixel matches in both the filter and in this part of the image. That's why we put **1** in the feature map.

What do you think happens when we hit the boundary of this image? What would you do? I'll show you what happens with these two diagrams:

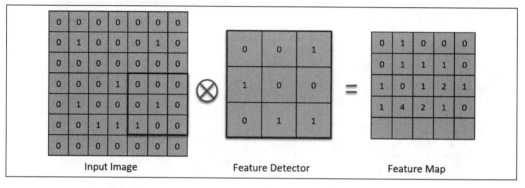

Figure 7: Step 1 – Convolution (3/5)

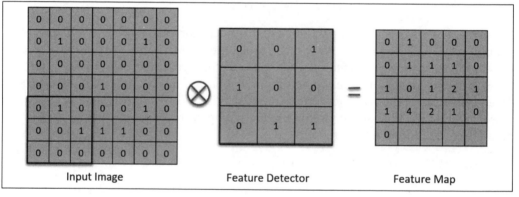

Figure 8: Step 1 – Convolution (4/5)

Here, we had this exact situation: in the first image, our filter hits the boundary. It turns out that our feature detector simply *jumps* to the next line.

The whole magic of the convolution wouldn't work if we had only one filter. In reality, we use many filters, which produce many different feature maps. This set of feature maps is called a **convolution layer**, or **convolutional layer**. Here's a diagram to recap:

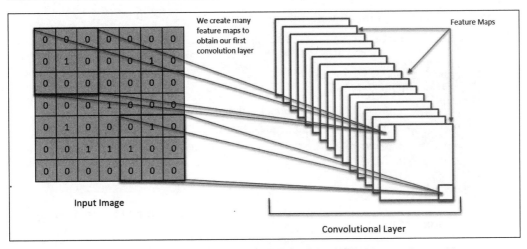

Figure 9: Step 1 – Convolution (5/5)

Here, we can see an input image to which many filters were applied. All together, they create a convolutional layer from many feature maps. This is the first step when building a CNN.

Now that we understand convolution, we can proceed to another important step — max pooling.

Step 2 – Max pooling

This step in CNNs is responsible for lowering the size of each feature map. When dealing with neural networks, we don't want to have too many inputs, otherwise our network wouldn't be able to learn properly because of the high complexity. Therefore, a method of reducing the size called **max pooling** needs to be introduced. It lets us reduce the size without losing any important features, and it makes features partially invariant to shifts (translations and rotations).

Technically, a max pooling algorithm is also based on an array sliding across the entire feature map. In this case, we are not searching for any features but, rather, for the maximum value in a specific area of a feature map.

Let me show you what I mean with this graphic:

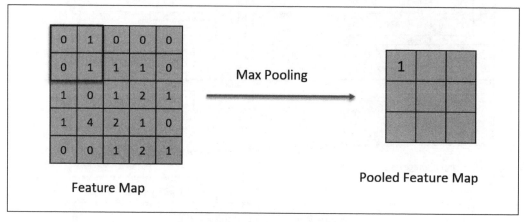

Figure 10: Step 2 – Max pooling (1/5)

In this example, we're taking the feature map, obtained after the convolution step we had before, and then we are running it through max pooling. As you can see, we have a window of size 2 x 2 looking for the highest values in the part of feature map it covers. In this case, it's 1.

Can you tell what will happen in the next iteration?

As you may have suspected, this window will slide to the right, although in a slightly different way than before. It moves like this:

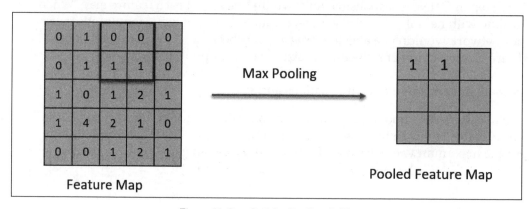

Figure 11: Step 2 – Max Pooling (2/5)

This window *jumps* its size to the right, which I hope you remember is different from the convolution step, where the feature detector slid one cell at a time. In this case, the highest value is 1 as well, and therefore we write **1** in the **pooled feature map**.

What happens this time when we hit the boundary of the feature map? Things look slightly different from before once again. This is what happens:

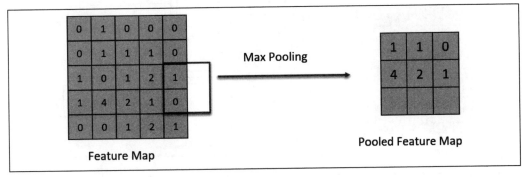

Figure 12: Step 2 – Max pooling (3/5)

The window crosses the boundary and searches for the highest value in the part of the feature map that is still inside the max pooling window. Yet again, the highest value is 1.

But what happens now? After all, there's no space left to go to the right. There's also only one row at the bottom left for max pooling. This is what the algorithm does:

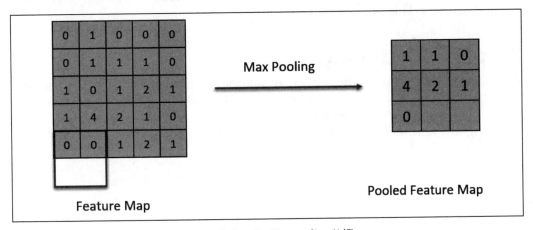

Figure 13: Step 2 – Max pooling (4/5)

As we can see, it once again crosses the boundary and searches for the highest value in what is inside the window. In this case, it is 0. This process is repeated until the window hits the bottom right corner of the feature map. To recap what our CNN looks like for now, have a look at the following diagram:

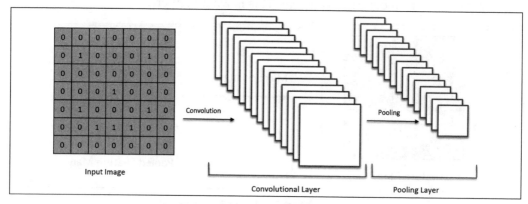

Figure 14: Step 2 – Max pooling (5/5)

We had a smiling face as input, then we ran it through convolution to obtain many feature maps, called the convolutional layer. Now we've run all the feature maps through max pooling and obtained many pooled feature maps, all together called the **pooling layer**.

Now we can continue to the next step, which will let us input the pooling layer into a neural network. This step is called **flattening**.

Step 3 – Flattening

This is a very short step. As the name may suggest, we change all the pooled feature maps from 2D arrays to 1D ones. As I mentioned before, this will let us input the image into a neural network with ease. So, how exactly will we achieve this? The following diagram should help you understand:

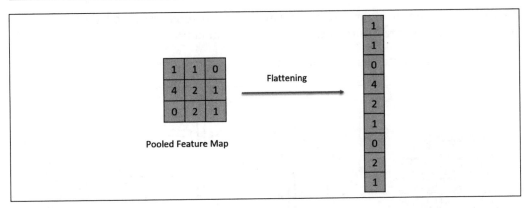

Figure 15: Step 3 – Flattening (1/3)

Here we go back to the pooled feature map we obtained before. To flatten it, we take pixel values starting from the top left, finishing at bottom right. An operation like this returns a 1D array, containing the same values as the 2D array we started with.

But remember, we don't have one pooled feature map, we have an entire layer of them. What do you think we should do with that?

The answer is simple: we put this entire layer into a single 1D flattened array, one pooled feature map after another. Why does it have to be 1D? This is because ANNs only accept 1D arrays as their inputs. All the layers in a traditional neural network are 1D, which means that the input has to be 1D as well. Therefore, we flatten all the pooled feature maps, like so:

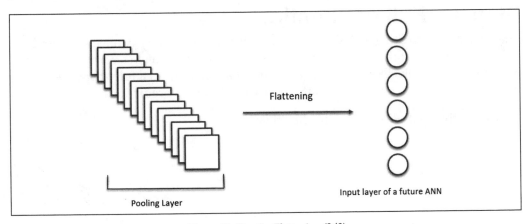

Figure 16: Step 3 – Flattening (2/3)

We've taken the entire layer and transformed it into a single flattened 1D array. We'll soon use this array as the input of a traditional neural network.

First, let's remind ourselves of what our model looks like now:

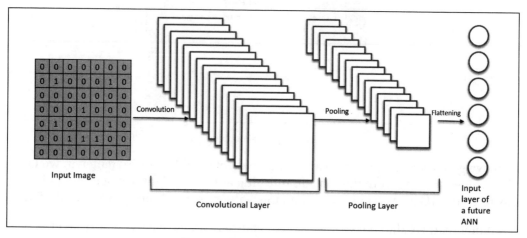

Figure 17: Step 3 – Flattening (3/3)

So, we have a Convolutional Layer, Pooling Layer, and a freshly added, flattened 1D layer. Now we can go back to a classic ANN, that is, a fully connected neural network, and treat this last layer as an input for this network. This leads us to the final step, **full connection**.

Step 4 – Full connection

The final step of creating a CNN is to connect it to a classic fully-connected neural network. Remember that we already have a 1D array telling us in a compressed way what the image looks like, so why not just use it as an input to a fully-connected neural network? After all, it's the latter that's able to make predictions.

That's exactly what we do next, just like this:

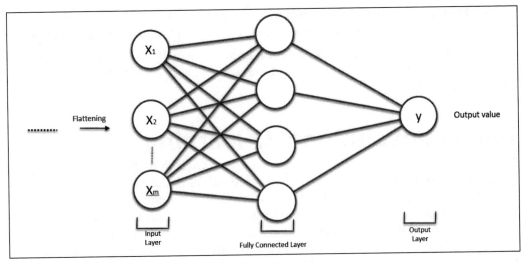

Figure 18: Step 4 – Full connection

After flattening, we input those returned values straight into the fully-connected neural network, which then yields the prediction—the output value.

You might be wondering how the back-propagation phase works now. In a CNN, back-propagation not only updates the weights in the fully-connected neural network, but also the filters used in the convolution step. The max pooling and flattening steps will remain the same, as there is nothing to update there.

In conclusion, CNNs look for some specific features. This is why they're mostly used when we are dealing with images, where searching for features is crucial. For example, when trying to recognize a sad and a happy face, a CNN needs to understand which mouth's shape means a sad face and which means a happy face. In order to obtain an output, a CNN has to run these steps:

1. **Convolution** – Applying filters to the input image. This operation will find the features our CNN is looking for and save them in a feature map.

2. **Max pooling** – Lowering the feature map size, by taking a maximum value in a given area and saving these values in a new array called pooled feature map.

3. **Flattening** – Changing the entire pooling layer (all pooled feature maps) to a 1D vector. This will allow us to input this vector into a neural network.

4. **Full connection** – Creating a neural network, which takes as input a flattened pooling layer and returns a value that we would like to predict. This last step lets us make predictions.

Deep convolutional Q-learning

In the chapter on deep Q-learning (*Chapter 9, Going Pro with Artificial Brains – Deep Q-Learning*), our inputs were vectors of encoded values defining the states of the environment. When working with images or videos, encoded vectors aren't the best inputs to describe a state (the input frame), simply because an encoded vector doesn't preserve the spatial structure of an image. The spatial structure is important because it gives us more information to help predict the next state, and predicting the next state is essential for our AI to learn the correct next move.

Therefore, we need to preserve the spatial structure. To do that, our inputs must be 3D images (2D for the array of pixels plus one additional dimension for the colors, as illustrated at the beginning of this chapter). For example, if we train an AI to play a video game, the inputs are simply the images of the screen itself, exactly what a human sees when playing the game.

Following this analogy, the AI acts like it has human eyes; it observes the input images on the screen when playing the game. Those input images go into a CNN (the eyes for a human), which detects the state in each image. Then they're forward-propagated through the pooling layers where max pooling is applied. Then the pooling layers are flattened into a 1D vector, which becomes the input of our deep Q-learning network (the exact same one as in *Chapter 9, Going Pro with Artificial Brains – Deep Q-Learning*). In the end, the same deep Q-learning process is run.

The following graph illustrates deep convolutional Q-learning applied to the famous game of Doom:

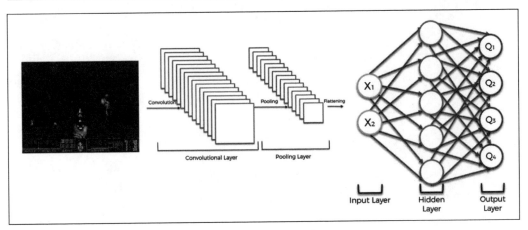

Figure 19: Deep convolutional Q-learning for Doom

In summary, deep convolutional Q-learning is the same as deep Q-learning, with the only differences being that the inputs are now images, and a CNN is added at the beginning of the fully-connected deep Q-learning network to detect the states of those images.

Summary

You've learned about another type of neural network—a Convolutional Neural Network.

We established that this network is used mostly with images and searches for certain features in these pictures. It uses three additional steps that ANNs don't have: convolution, where we search for features; max pooling, where we shrink the image in size; and flattening, where we flatten 2D images to a 1D vector so that we can input it into a neural network.

In the next chapter, you'll build a deep convolutional Q-learning model to solve a classic gaming problem: Snake.

13
AI for Games – Become the Master at Snake

This is the last practical chapter; congratulations on finishing the previous ones! I hope you really enjoyed them. Now, let's leave aside business problems and self-driving cars. Let's have some fun by playing a popular game called Snake and making an AI that teaches itself to play this game!

That's exactly what we'll do in this chapter. The model we'll implement is called deep convolutional Q-learning, using a **Convolutional Neural Network (CNN)**.

Our AI won't be perfect, and it won't fill in the entire map, but after some training it will start playing at a level comparable with humans.

Let's start tackling this problem by looking at what the game looks like and what the target is.

Problem to solve

First, let's have a look at the game itself:

Figure 1: The Snake game

Does that look somewhat familiar to you?

I'm pretty convinced that it will; everyone's played Snake at least once in their life.

The game is pretty simple; it consists of a snake and an apple. We control the snake and our aim is to eat as many apples as possible.

Sounds easy? Well, there's a small catch. Every time our snake eats an apple, our snake gets larger by one tile. This means that the game is unbelievably simple at the beginning, but it gets gradually harder, to the point where it becomes a strategic game.

Also, when controlling our snake, we can't hit ourselves, nor the borders of the board. This rather predictably results in us losing.

Now that we understand the problem, we can progress to the first step when creating an AI – building the environment!

Building the environment

This time, as opposed to some of the other practical sections in this book, we don't have to specify any variables or make any assumptions. We can just go straight to the three crucial steps present in every deep Q-learning project:

1. Defining the states

2. Defining the actions

3. Defining the rewards

Let's begin!

Defining the states

In every previous example, our states were a 1D vector that represented some values that define the environment. For example, for our self-driving car we had the information gathered from the three sensors around the car and the car's position. All of these were put into a single 1D array.

But what if we want to make something slightly more realistic? What if we want the AI to see and gather information from the same source as we do? Well, that's what we'll do in this chapter. Our AI will see exactly the same board as we see when playing Snake!

The state of the game should be a 2D array representing the board of the game, exactly the same thing that we can see.

There's just one problem with this solution. Take a look at the following image, and see if you can answer the question: which way is our snake moving right now?

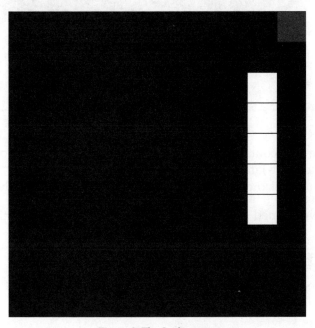

Figure 2: The Snake game

If you said "I don't know," then you're exactly right.

Based on a single frame, we can't tell which way our snake is going. Therefore, we'll need to stack multiple images, and then input all of them at once to a Convolutional Neural Network. This will result in us having 3D states rather than 2D ones.

So, just to recap:

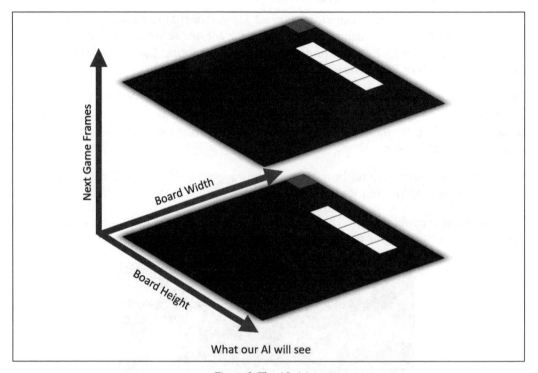

Figure 3: The AI vision

We'll have a 3D array, containing next game frames stacked on top of each other, where the top one is the latest frame obtained from our game. Now, we can clearly see which way our AI is moving; in this case it's going up, toward the apple.

Now that we have defined states, we can go the next step: defining the actions!

Defining the actions

When we play Snake on a phone or a website, there are four actions available for us to take:

1.   Go up

2.   Go down

3. Go right

4. Go left

However, if the action we take would require the snake to make a 180° turn directly back on itself, then the game blocks this action and the snake continues going in its current direction.

In the preceding example, if we were to select action 2 – go down–our snake would still continue going up, because going down is impossible as the snake can't make a 180° turn directly back on itself.

It's worth noting that all of these actions are relative to the board, not the snake; they're not affected by the current movement of the snake. Going up, down, right, or left always means going up, down, right, or left with respect to the board, not to the snake's current direction of movement.

Alright, so right now you might be in one of these two groups when it comes to deciding what actions we model in our AI:

1. We can use these four same actions for our AI.

2. We can't use these same actions, because blocking certain moves will be confusing for our AI. Instead, we should invent a way to tell the snake to go left, go right, or keep going.

We actually can use these same actions for our AI!

Why won't it be confusing for our agent? That's because as long as our AI agent gets rewards for the actions it chose, and not for the action ultimately performed by the snake, then deep Q-learning will work and our AI will understand that in the example above choosing either *go up* or *go down* results in the same outcome.

For example, let's say that the AI-controlled snake is currently going left. It chooses action 3, go right; and because that would cause the snake to make a 180° turn back on itself, instead the snake continues going left. Let's say that action means the snake crashes into the wall and, as a result, dies. In order for this not to be confusing for our agent, all we need to do is tell it that the action of *go right* caused it to crash, even though the snake kept moving left.

Think of it as teaching an AI to play with the actual buttons on a phone. If you keep trying to make your snake double back on itself when it's moving left, by pressing the go right button over and over again, the game will keep ignoring the impossible move you keep telling it to do, keep going left, and eventually crash. That's all the AI needs to learn.

This is because, remember, in deep Q-learning we only update the Q-values of the action that the AI takes. If our snake is going left, and the AI decides to go right and the snake dies, it needs to understand that the action of *go right* caused it to get the negative reward, not the fact that the snake moved left; even though choosing the action *go left* would cause the same outcome.

I hope you understand that the AI can use the same actions as we use when we play. We can continue to the next, final step – defining the rewards!

Defining the rewards

This last step is pretty simple; we just need three rewards:

1. Reward for eating an apple
2. Reward for dying
3. The living penalty

The first two are hopefully easy to understand. After all, we want to encourage our agent to eat as many apples as possible and therefore we will set its reward to be positive. To be precise: **eating an apple = +2**

Meanwhile, we want to discourage our snake from dying. That's why we set that reward to be a negative one. To be precise: **dying = -1**

Then comes the final reward: the living penalty.

What is that, and why is it necessary? We have to convince our agent that collecting apples as quickly as possible, without dying, is a good idea. If we were to only have the two rewards we've already defined, our agent would simply travel around the entire map, hoping that at some point it finds an apple. It wouldn't understand that it needs to collect apples as quickly as it can.

That's why we introduce the living penalty. It will slightly punish our AI for every action it takes, unless this action leads to dying or collecting an apple. This will show our agent that it needs to collect apples quickly, as only moves that collect an apple lead to gaining a positive reward. So, how big this reward should be? Well, we don't want to punish it too much. To be precise: **living penalty =-0.03**

If you want to tinker with these rewards, the absolute value of this reward should always be relatively small compared to the other rewards, for dying (-1) and collecting an apple (+2).

AI solution

As always, the AI solution for deep Q-learning consists of two parts:

1. **Brain** – the neural network that will learn and take actions
2. **Experience replay memory** – the memory that will store our experience; the neural network will learn from this memory

Let's tackle those now!

The brain

This part of the AI solution will be responsible for teaching, storing, and evaluating our neural network. To build it, we're going to use a CNN!

Why a CNN? When explaining the theory behind them, I mentioned that they're often used when "our environment as state returns images," and that's exactly what we're dealing with here. We've already established that the game state is going to be a stacked 3D array containing the last few game frames.

In the previous chapter, we discussed that a CNN takes a 2D image as input, not a stacked 3D array of images; but do you remember this graphic?

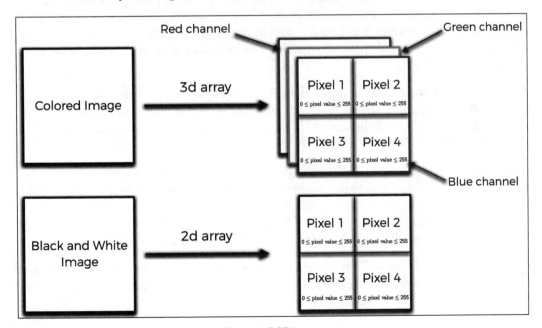

Figure 4: RGB images

Here, I informed you that the RGB images are represented by 3D arrays that contain every single 2D channel of this image. Does that sound familiar? We can use the very same method for our problem. Just like each color in the RGB structure, we'll simply input every game frame as a new channel, which will give us a 3D array, which we will be able to input into a CNN.

In reality, CNNs usually only support 3D arrays as inputs. In order to input a 2D array, you need to create a fake single channel that transforms a 2D array into a 3D one.

When it comes to the CNN architecture, we'll have two convolution layers separated by a pooling layer. One convolution layer will have 32 3x3 filters, and the other one will have 64 2x2 filters. The pooling layer will shrink the size by 2, as the pooling window size will be 2x2. Why such an architecture? It's a classic one, found in many research papers, which I arbitrarily chose as common practice and which turned out to work brilliantly.

Our neural network will have one hidden layer with 256 neurons, and an output layer with 4 neurons; one for each of our possible outcome actions.

We also need to set two last parameters for our CNN – learning rate and input shape.

Learning rate, which was used in the previous examples, is a parameter that specifies by how much we update the weights in the neural network. Too small and it won't learn, too big and it won't learn for a different reason; the changes will be too big for any optimization. I found through experimentation that a good learning rate for this example is 0.0001.

We've already agreed that the input should be a 3D array containing last frames obtained from our game. To be exact, we will not be reading pixels from our screen. Instead, we'll read the direct 2D array that represents our game's screen at a particular time.

As you've probably noticed, our game is built on a grid. In the example we are using, the grid is 10x10. Then, inside the environment is an array with the same size (10x10), telling us mathematically what the board looks like. For example, if we have part of the snake in one cell, then we place the value 0.5 in the corresponding cell in our 2D array, which we will read. An apple is described as value 1 in this array.

Now that we know how we'll see one frame, we need to decide how many previous frames we'll use when we describe the current game state. 2 should be enough, since we can discern from that which way the snake is going, but to make sure, we'll have 4.

Can you tell me exactly what shape our input to the CNN will be?

It'll be 10x10x4, which gives us a 3D array!

The experience replay memory

As defined in the theoretical chapter of deep Q-learning, we need to have a memory that stores experience gathered during training.

We'll store the following data:

- Current state – The game state the AI was in when it performed an action (what we inputted to our CNN)
- Action – Which action was undertaken
- Reward – The reward gained by performing this action on the current state
- Next state – What happened (how the state looked) after performing the action
- Game over – Information about whether we have lost or not

Also, we always have to specify two parameters for every experience replay memory:

- Memory size – The maximum size of our memory
- Gamma – The discount factor, existent in the Bellman equation

We'll set the memory size to 60,000 and the gamma parameter to 0.9.

There's one last thing to specify here.

I told you that our AI will learn from this memory, and that's true; but the AI won't be learning from the entire memory. Rather, it will learn from a small batch taken from it. The parameter that specifies this size will be called batch size, and in this example, we'll set its value to 32. That means that our AI will learn every iteration from a batch of this size taken from experience replay memory.

Now that you understand everything you have to code, you can get started!

Implementation

You'll implement the entire AI code and the Snake game in five files:

1. `environment.py` file – The file containing the environment (Snake game)
2. `brain.py` file – The file in which we build our CNN

3. DQN.py – The file that builds the Experience Replay Memory

4. train.py – The file where we will train our AI to play Snake

5. test.py – The file where we will test our AI to see how well it performs

You can find all of them on the GitHub page along with a pre-trained model. To get there, select Chapter 13 folder on the main page.

We'll go through each file in the same order. Let's start building the environment!

Step 1 – Building the environment

Start this first, important step by importing the libraries you'll need. Like this:

```
4      # Importing the libraries
5      import numpy as np
6      import pygame as pg
```

You'll only use two libraries: NumPy and PyGame. The former is really useful when dealing with lists or arrays, and the latter will be used to build the entire game – to draw the snake and the apple, and update the screen.

Now, let's create the Environment class which will contain all the information, variables and methods that you need for your game. Why a class? This is because it makes things easier for you later on. You'll be able to call specific methods or variables from the object of this class.

The first method that you always have to have is the __init__ method, always called when a new object of this class is created in the main code. To create this class along with this __init__ method, you need to write:

```
8      # Initializing the Environment class
9      class Environment():
10
11         def __init__(self, waitTime):
12
13             # Defining the parameters
14             self.width = 880              # width of the game window
15             self.height = 880             # height of the game window
16             self.nRows = 10              # number of rows in our board
17             self.nColumns = 10           # number of columns in our
       board
18             self.initSnakeLen = 2        # initial length of the snake
```

```
19              self.defReward = -0.03      # reward for taking an action
        - The Living Penalty
20              self.negReward = -1.        # reward for dying
21              self.posReward = 2.         # reward for collecting an
        apple
22              self.waitTime = waitTime    # slowdown after taking an
        action
23
24              if self.initSnakeLen > self.nRows / 2:
25                  self.initSnakeLen = int(self.nRows / 2)
26
27              self.screen = pg.display.set_mode((self.width, self.
        height))
28
29              self.snakePos = list()
30
31              # Creating the array that contains mathematical
        representation of the game's board
32              self.screenMap = np.zeros((self.nRows, self.nColumns))
33
34              for i in range(self.initSnakeLen):
35                  self.snakePos.append((int(self.nRows / 2) + i,
        int(self.nColumns / 2)))
36                  self.screenMap[int(self.nRows / 2) + i][int(self.
        nColumns / 2)] = 0.5
37
38              self.applePos = self.placeApple()
39
40              self.drawScreen()
41
42              self.collected = False
43              self.lastMove = 0
```

You create a new class, the Environment() class, along with its __init__ method.
This method only takes one argument, which is waitTime. Then after defining the
method, create a list of constants, each of which is explained in the inline comments.
After that, you perform some initialization. You make sure the snake is half the
length of the screen or less on lines 24 and 25, and set the screen up on line 27. One
important thing to note is that you create the screenMap array on line 32, which
represents the board more mathematically. 0.5 in a cell means that this cell is taken
by the snake, and 1 in a cell means that this cell is taken by the apple.

On lines 34 to 36, you place the snake in the middle of the screen, facing upward, and then in the remaining lines you place an apple using the `placeapple()` method (which we are about to define), draw the screen, set that the apple hasn't been collected, and set that there's no last move.

That's the very first method completed. Now you can proceed to the next one:

```
# Building a method that gets new, random position of an apple
def placeApple(self):
    posx = np.random.randint(0, self.nColumns)
    posy = np.random.randint(0, self.nRows)
    while self.screenMap[posy][posx] == 0.5:
        posx = np.random.randint(0, self.nColumns)
        posy = np.random.randint(0, self.nRows)

    self.screenMap[posy][posx] = 1

    return (posy, posx)
```

This short method places an apple in a new, random spot in your `screenMap` array. You'll need this method when our snake collects the apple and a new apple needs to be placed. It also returns the random position of the new apple.

Then, you'll need a function that draws everything for you to see:

```
# Making a function that draws everything for us to see
def drawScreen(self):

    self.screen.fill((0, 0, 0))

    cellWidth = self.width / self.nColumns
    cellHeight = self.height / self.nRows

    for i in range(self.nRows):
        for j in range(self.nColumns):
            if self.screenMap[i][j] == 0.5:
                pg.draw.rect(self.screen, (255, 255, 255),
    (j*cellWidth + 1, i*cellHeight + 1, cellWidth - 2, cellHeight - 2))
            elif self.screenMap[i][j] == 1:
                pg.draw.rect(self.screen, (255, 0, 0),
    (j*cellWidth + 1, i*cellHeight + 1, cellWidth - 2, cellHeight - 2))

    pg.display.flip()
```

As you can see, the name of this method is `drawScreen` and it doesn't take any arguments. Here you simply empty the entire screen, then fill it in with white tiles where the snake is and with a red tile where the apple is. At the end, you update the screen with `pg.display.flip()`.

Now, you need a function that will update the snake's position and not the entire environment:

```
# A method that updates the snake's position
def moveSnake(self, nextPos, col):

    self.snakePos.insert(0, nextPos)

    if not col:
        self.snakePos.pop(len(self.snakePos) - 1)

    self.screenMap = np.zeros((self.nRows, self.nColumns))

    for i in range(len(self.snakePos)):
        self.screenMap[self.snakePos[i][0]][self.snakePos[i][1]] =
0.5

    if col:
        self.applePos = self.placeApple()
        self.collected = True

    self.screenMap[self.applePos[0]][self.applePos[1]] = 1
```

You can see that this new method takes two arguments: `nextPos` and `col`. The former tells you where the head of the snake will be after performing a certain action. The latter will inform you whether the snake has collected an apple by taking this action, or not. Remember that if the snake has collected an apple, then the length of the snake increases by 1. If you go deep into this code, you can see that, but we won't go into detail here since it's not so relevant for the AI. You can also see that if the snake has collected an apple, a new one is spawned in a new spot.

Now, let's move on to the most important part of this code. You define a function that will update the entire environment. It will move your snake, calculate the reward, check if you lost, and return a new game frame. This is how it starts:

```
# The main method that updates the environment
def step(self, action):
    # action = 0 -> up
    # action = 1 -> down
    # action = 2 -> right
```

```
        # action = 3 -> left

        # Resetting these parameters and setting the reward to the
living penalty
        gameOver = False
        reward = self.defReward
        self.collected = False

        for event in pg.event.get():
            if event.type == pg.QUIT:
                return

        snakeX = self.snakePos[0][1]
        snakeY = self.snakePos[0][0]

        # Checking if an action is playable and if not then it is
changed to the playable one
        if action == 1 and self.lastMove == 0:
            action = 0
        if action == 0 and self.lastMove == 1:
            action = 1
        if action == 3 and self.lastMove == 2:
            action = 2
        if action == 2 and self.lastMove == 3:
            action = 3
```

As you can see, this method is called step and it takes one argument: the action that tells you which way you want the snake to be going. Just beneath the method's definition, in the comments, you can see which action means which direction.

Then you reset some variables. You set gameOver to False as this bool variable will tell you if you lost after performing this action. You set reward to defReward, as this is the living penalty; it can change if we collect an apple or die later.

Then there's a for loop. It's there to make sure the PyGame window doesn't freeze; this is a requirement of the PyGame library. It just has to be there.

snakeX and snakeY tell you what the head position of the snake is. It'll be used by the algorithm later, to determine what happens after the head moves.

In the last few lines, you can see the algorithm that blocks impossible actions. Just to recap, an impossible action is the one that requires the snake to make a 180° turn in place. lastMove tells you which way the snake is going right now, and is compared with action. If these lead to a contradiction, then action is set to lastMove.

Still inside this method, you update the snake position, check for game over, and calculate the reward, like so:

```
# Checking what happens when we take this action
if action == 0:
    if snakeY > 0:
        if self.screenMap[snakeY - 1][snakeX] == 0.5:
            gameOver = True
            reward = self.negReward
        elif self.screenMap[snakeY - 1][snakeX] == 1:
            reward = self.posReward
            self.moveSnake((snakeY - 1, snakeX), True)
        elif self.screenMap[snakeY - 1][snakeX] == 0:
            self.moveSnake((snakeY - 1, snakeX), False)
    else:
        gameOver = True
        reward = self.negReward
```

Here you check what happens if the snake goes up. If the head of the snake is already in the top row (row no. 0) then you've obviously lost, since the snake hits the wall. So, reward is set to negReward and gameOver is set to True. Otherwise, you check what lies ahead of the snake.

If the cell ahead already contains part of the snake's body, then you've lost. You check that in the first if statement, then set gameOver to True and reward to negReward.

Else if the cell ahead is an apple, then you set reward to posReward. You also update the snake's position by calling the method you created just before this one.

Else if the cell ahead is empty, then you don't update reward in any way. You call the same method again, but this time with the col argument set to False, since the snake hasn't collected an apple. You go through the same process for every other action. I won't go through every line, but have a look at the code:

```
elif action == 1:
    if snakeY < self.nRows - 1:
        if self.screenMap[snakeY + 1][snakeX] == 0.5:
            gameOver = True
            reward = self.negReward
        elif self.screenMap[snakeY + 1][snakeX] == 1:
            reward = self.posReward
            self.moveSnake((snakeY + 1, snakeX), True)
        elif self.screenMap[snakeY + 1][snakeX] == 0:
            self.moveSnake((snakeY + 1, snakeX), False)
    else:
```

```
            gameOver = True
            reward = self.negReward

        elif action == 2:
            if snakeX < self.nColumns - 1:
                if self.screenMap[snakeY][snakeX + 1] == 0.5:
                    gameOver = True
                    reward = self.negReward
                elif self.screenMap[snakeY][snakeX + 1] == 1:
                    reward = self.posReward
                    self.moveSnake((snakeY, snakeX + 1), True)
                elif self.screenMap[snakeY][snakeX + 1] == 0:
                    self.moveSnake((snakeY, snakeX + 1), False)
            else:
                gameOver = True
                reward = self.negReward

        elif action == 3:
            if snakeX > 0:
                if self.screenMap[snakeY][snakeX - 1] == 0.5:
                    gameOver = True
                    reward = self.negReward
                elif self.screenMap[snakeY][snakeX - 1] == 1:
                    reward = self.posReward
                    self.moveSnake((snakeY, snakeX - 1), True)
                elif self.screenMap[snakeY][snakeX - 1] == 0:
                    self.moveSnake((snakeY, snakeX - 1), False)
            else:
                gameOver = True
                reward = self.negReward
```

Simply handle every single action in the same way you did with the action of going up. Check if the snake didn't hit the walls, check what lies ahead of the snake and update the snake's position, `reward`, and `gameOver` accordingly.

There are two more steps in this method; let's jump straight into the first one:

```
        # Drawing the screen, updating last move and waiting the wait
    time specified
        self.drawScreen()

        self.lastMove = action

        pg.time.wait(self.waitTime)
```

You update our screen by drawing the snake and the apple on it, then change `lastMove` to `action`, since your snake has already moved and now it's moving in the `action` direction.

The last step in this method is to return what the game looks like now, what the reward is that was obtained, and whether you've lost, like this:

```
        # Returning the new frame of the game, the reward obtained
    and whether the game has ended or not
        return self.screenMap, reward, gameOver
```

`screenMap` gives you the information you need about what the game looks like after performing an action, `reward` gives you the collected reward from taking this action, and `gameOver` tells you whether you lost or not.

That's it for this method! To have a complete `Environment` class, you only need to make a function that will reset the environment, like this `reset` method:

```
        # Making a function that resets the environment
    def reset(self):
        self.screenMap  = np.zeros((self.nRows, self.nColumns))
        self.snakePos = list()

        for i in range(self.initSnakeLen):
            self.snakePos.append((int(self.nRows / 2) + i, int(self.
    nColumns / 2)))
            self.screenMap[int(self.nRows / 2) + i][int(self.nColumns
    / 2)] = 0.5

        self.screenMap[self.applePos[0]][self.applePos[1]] = 1

        self.lastMove = 0
```

It simply resets the game board (`screenMap`), as well as the snake's position, to the default, which is the middle of the board. It also sets the apple's position to the same as it was in the last round.

Congratulations! You've just finished building the environment. Now, we'll proceed to the second step, building the brain.

Step 2 – Building the brain

This is where you'll build our brain with a Convolutional Neural Network. You'll also set some parameters for its training and define a method that loads a pre-trained model for testing.

Let's begin!

As always, you start by importing the libraries that you'll use, like this:

```
# Importing the libraries
import keras
from keras.models import Sequential, load_model
from keras.layers import Dense, Dropout, Conv2D, MaxPooling2D, Flatten
from keras.optimizers import Adam
```

As you've probably noticed, all of the classes are a part of the Keras library, which is the one you're going to use in this chapter. Keras is actually the only library that you'll use in this file. Let's go through each of these classes and methods right now:

1. Sequential – A class that allows you to initialize a neural network, and defines the general structure of this network.

2. load_model – A function that loads a model from a file.

3. Dense – A class to create fully connected layers in an Artificial Neural Network (ANN).

4. Dropout – A class that adds dropout to our network. You've seen it used already, in *Chapter 8, AI for Logistics – Robots in a Warehouse*.

5. Conv2D – A class that builds convolution layers.

6. MaxPooling2D – A class that builds max pooling layers.

7. Flatten – A class that performs flattening, so that you'll have an input for a classic ANN.

8. Adam – An optimizer, which will optimize your neural network. It's used when training the CNN.

Now you've imported your library, you can continue by creating a class called Brain, where all these classes and methods are used. Start by defining a class and the __init__ method, like this:

```
# Creating the Brain class
class Brain():

    def __init__(self, iS = (100,100,3), lr = 0.0005):

        self.learningRate = lr
        self.inputShape = iS
        self.numOutputs = 4
        self.model = Sequential()
```

You can see that the __init__ method takes two arguments: iS (input shape) and lr (learning rate). Then you define some variables that will be associated with this class: learningRate, inputShape, numOutputs. Set numOutputs to 4, as this is how many actions our AI can take. Then, in the last line, create an empty model. To do this, use the Sequential class, which we imported earlier.

Doing this will allow you to add all the layers that you need to the model. That's exactly what you do with these lines:

```
20          # Adding layers to the model
21          self.model.add(Conv2D(32, (3,3), activation = 'relu',
       input_shape = self.inputShape))
22
23          self.model.add(MaxPooling2D((2,2)))
24
25          self.model.add(Conv2D(64, (2,2), activation = 'relu'))
26
27          self.model.add(Flatten())
28
29          self.model.add(Dense(units = 256, activation = 'relu'))
30
31          self.model.add(Dense(units = self.numOutputs))
```

Let's break this code down into lines:

Line 21: You add a new convolution layer to your model. It has 32 3x3 filters with the ReLU activation function. You need to specify the input shape here as well. Remember that the input shape is one of the arguments of this function, and is saved under the inputShape variable.

Line 23: You add a max pooling layer. The window's size is 2x2, which will shrink our feature maps in size by 2.

Line 25: You add the second convolution layer. This time it has 64 2x2 filters, with the same ReLU activation function. Why ReLU this time? I tried some other activation functions experimentally, and it turned out that for this AI ReLU worked the best.

Line 27: Having applied convolution, you receive new feature maps, which you flatten to a 1D vector. That's exactly what this line does – it flattens 2D images to a 1D vector, which you'll then be able to use as the input to your neural network.

Line 29: Now, you're in the full connection step – you're building the traditional ANN. This specific line adds a new hidden layer with `256` neurons and the ReLU activation function to our model.

Line 31: You create the last layer in your neural network – the output layer. How big is it? Well, it has to have as many neurons as there are actions that you can take. You put that value under the `numOutputs` variable earlier, and the value is equal to `4`. You don't specify the activation function here, which means that the activation function will be linear as a default. It turns out that in this case, during training, using a linear output works better than a Softmax output; it makes the training more efficient.

You also have to `compile` your model. This will tell your code how to calculate the error, and which optimizer to use when training your model. You can do it with this single line:

```
# Compiling the model
self.model.compile(loss = 'mean_squared_error', optimizer =
Adam(lr = self.learningRate))
```

Here, you use a method that's a part of the `Sequential` class (that's why you can use your model to call it) to do just that. The method is called `compile` and, in this case, takes two arguments. `loss` is a function that tells the AI how to calculate the error of your neural network; you'll use `mean_squared_error`. The second parameter is the optimizer. You've already imported the `Adam` optimizer, and you use it here. The learning rate for this optimizer was one of the arguments of the `__init__` method of this class, and its value is represented by the `learningRate` variable.

There's only one step left to do in this class – make a function that will load a model from a file. You do it with this code:

```
# Making a function that will load a model from a file
def loadModel(self, filepath):
    self.model = load_model(filepath)
    return self.model
```

You can see that you've created a new function called `loadModel`, which takes one argument – `filepath`. This parameter is the file path to the pre-trained model. Once you've defined the function, you can actually load the model from this file path. To do so, you use the `load_model` method, which you imported earlier. This function takes the same argument – `filepath`. Then in the final line, you return the loaded model.

Congratulations! You've just finished building the brain.

Let's advance on our path, and build the experience replay memory.

Step 3 – Building the experience replay memory

You'll build this memory now, and later, you'll train your model from small batches of this memory. The memory will contain information about the game state before taking the action, the action that was taken, the reward gained, and the game state after performing the action.

I have some excellent news for you – do you remember this code?

```
# AI for Games - Beat the Snake game
# Implementing Deep Q-Learning with Experience Replay

# Importing the libraries
import numpy as np


# IMPLEMENTING DEEP Q-LEARNING WITH EXPERIENCE REPLAY

class Dqn(object):

    # INTRODUCING AND INITIALIZING ALL THE PARAMETERS AND VARIABLES
OF THE DQN
    def __init__(self, max_memory = 100, discount = 0.9):
        self.memory = list()
        self.max_memory = max_memory
        self.discount = discount

    # MAKING A METHOD THAT BUILDS THE MEMORY IN EXPERIENCE REPLAY
    def remember(self, transition, game_over):
        self.memory.append([transition, game_over])
        if len(self.memory) > self.max_memory:
            del self.memory[0]

    # MAKING A METHOD THAT BUILDS TWO BATCHES OF INPUTS AND TARGETS BY
EXTRACTING TRANSITIONS FROM THE MEMORY
    def get_batch(self, model, batch_size = 10):
        len_memory = len(self.memory)
        num_inputs = self.memory[0][0][0].shape[1]
        num_outputs = model.output_shape[-1]
        inputs = np.zeros((min(len_memory, batch_size), num_inputs))
        targets = np.zeros((min(len_memory, batch_size), num_outputs))
        for i, idx in enumerate(np.random.randint(0, len_memory, size
= min(len_memory, batch_size))):
            current_state, action, reward, next_state = self.
memory[idx][0]
```

```
            game_over = self.memory[idx][1]
            inputs[i] = current_state
            targets[i] = model.predict(current_state)[0]
            Q_sa = np.max(model.predict(next_state)[0])
            if game_over:
                targets[i, action] = reward
            else:
                targets[i, action] = reward + self.discount * Q_sa
        return inputs, targets
```

You'll use almost the same code, with only two small changes.

First, you get rid of this line:

```
        num_inputs = self.memory[0][0][0].shape[1]
```

And then change this line:

```
        inputs = np.zeros((min(len_memory, batch_size), num_inputs))
```

To this one:

```
        inputs = np.zeros((min(len_memory, batch_size), self.memory[0]
    [0][0].shape[1],self.memory[0][0][0].shape[2],self.memory[0][0][0].
    shape[3]))
```

Why did you have to do this? Well, you got rid of the first line since you no longer have a 1D vector of inputs. Now you have a 3D array.

Then, if you look closely, you'll see that you didn't actually change inputs. Before, you had a 2D array, one dimension of which was batch size and the other of which was number of inputs. Now, things are very similar; the first dimension is once again the batch size, and the last three correspond to the size of the input as well!

Since our input is now a 3D array, you wrote .shape[1], .shape[2], and .shape[3]. What exactly are those shapes?

.shape[1] is the number of rows in the game (in your case 10). .shape[2] is the number of columns in the game (in your case 10). .shape[3] is the number of last frames stacked onto each other (in your case 4).

As you can see, you didn't really change anything. You just made the code work for our 3D inputs.

I also renamed this dqn.py file to DQN.py and renamed the class DQN to Dqn.

That's that! That was probably much simpler than most of you expected it to be.

You can finally start training your model. We'll do that in the next section – training the AI.

Step 4 – Training the AI

This is, by far, the most important step. Here we finally teach our AI to play Snake!

As always, start by importing the libraries you need:

```
# Importing the libraries
from environment import Environment
from brain import Brain
from DQN import Dqn
import numpy as np
import matplotlib.pyplot as plt
```

In the first three lines you import the tools that you created earlier, including the `Brain`, the `Environment`, and the experience replay memory.

Then, in the following two lines, you import the libraries that you'll use. These include NumPy and Matplotlib. You'll already recognize the former; the latter will be used to display your model's performance. To be specific, it will help you display a graph that, every 100 games, will show you the average number of apples collected.

That's all for this step. Now, define some hyperparameters for your code:

```
# Defining the parameters
memSize = 60000
batchSize = 32
learningRate = 0.0001
gamma = 0.9
nLastStates = 4

epsilon = 1.
epsilonDecayRate = 0.0002
minEpsilon = 0.05

filepathToSave = 'model2.h5'
```

I'll explain them in this list:

1. `memSize` – The maximum size of your experience replay memory.
2. `batchSize` – The size of the batch of inputs and targets that you get at each iteration from your experience replay memory for your model to train on.

3. `learningRate` – The learning rate for your `Adam` optimizer in the `Brain`.

4. `gamma` – The discount factor for your experience replay memory.

5. `nLastStates` – How many last frames you save as your current state of the game. Remember, you'll input a 3D array of size `nRows x nColumns x nLastStates` to your CNN in the `Brain`.

6. `epsilon` – The initial epsilon, the chance of taking a random action.

7. `epsilonDecayRate` – By how much you decrease `epsilon` after every single game/epoch.

8. `minEpsilon` – The lowest possible epsilon, after which it can't be adjusted any lower.

9. `filepathToSave` – Where you want to save your model.

There you go – you've defined the hyperparameters. You'll use them later when you write the rest of the code. Now, you have to create an environment, a brain, and an experience replay memory:

```
# Creating the Environment, the Brain and the Experience Replay Memory
env = Environment(0)
brain = Brain((env.nRows, env.nColumns, nLastStates), learningRate)
model = brain.model
dqn = Dqn(memSize, gamma)
```

You can see that in the first line you create an object of the `Environment` class. You need to specify one variable here, which is the slowdown of your environment (wait time between moves). You don't want any slowdown during the training, so you input 0 here.

In the next line you create an object of the `Brain` class. It takes two arguments – the input shape and the learning rate. As I've mentioned multiple times, the input shape will be a 3D array of size `nRows x nColumns x nLastStates`, so that's what you type in here. The second argument is the learning rate, and since you've created a variable for that, you simply input the name of this variable – `learningRate`. After this line you take the model of this `Brain` class and create an instance of this model in your code. Keep things simple, and call it `model`.

In the last line you create an object of the `Dqn` class. It takes two arguments – the maximum size of the memory, and the discount factor for the memory. You've specified two variables, `memSize` and `gamma`, for just that, so you use them here.

Now, you need to write a function that will reset the states for your AI. You need it because the states are quite complicated, and resetting them in the main code would mess it up a lot. Here's what it looks like:

```
30      # Making a function that will initialize game states
31      def resetStates():
32          currentState = np.zeros((1, env.nRows, env.nColumns,
        nLastStates))
33
34          for i in range(nLastStates):
35              currentState[:,:,:,i] = env.screenMap
36
37          return currentState, currentState
```

Let's break it down into separate lines:

Line 31: You define a new function called resetStates. It doesn't take any arguments.

Line 32: You create a new array called currentState. It's full of zeros, but you may ask why it's 4D; shouldn't the input be 3D as we said? You're absolutely right, and it will be. The first dimension is called batch size and simply says how many inputs you input to your neural network at once. You'll only input one array at a time, so the first size is 1. The next three sizes correspond to the size of the input.

Lines 34-35: In a for loop, which will be executed nLastStates times, you set the board for each layer in your 3D state to the current, initial look of the game board from your environment. Every frame in your state will look the same initially, the same way the board of the game looks when you start a game.

Line 37: This function will return two currentStates. Why? This is because you need two game state arrays. One to represent the board before you've taken an action, and one to represent the board after you've taken an action.

Now you can start writing the code for the entire training. First, create a couple of useful variables, like this:

```
# Starting the main loop
epoch = 0
scores = list()
maxNCollected = 0
nCollected = 0.
totNCollected = 0
```

epoch will tell you which epoch/game you're in right now. scores is a list in which you save the average scores per game after every 100 games/epochs. maxNCollected tells you the highest score obtained so far in the training, while nCollected is the score in each game/epoch. The last variable, totNCollected, tells you how many apples you've collected over 100 epochs/games.

Now you start an important, infinite `while` loop, like this:

```
while True:
    # Resetting the environment and game states
    env.reset()
    currentState, nextState = resetStates()
    epoch += 1
    gameOver = False
```

Here, you iterate through every game, every epoch. That's why you restart the environment in the first line, create new `currentState` and `nextState` in the next line, increase `epoch` by one, and set `gameOver` to `False` as you obviously haven't lost yet.

Note that this loop doesn't end; therefore, the training never stops. We do it this way because we don't have a set goal for when to stop the training, since we haven't defined what a satisfactory result for our AI would be. We could calculate the average result, or a similar metric, but then training might take too long. I prefer to keep the training going and you can just stop the training whenever you want. A good time to stop is when the AI reaches an average of six apples per game, or you can even go up to 12 apples per game if you want better performance.

You've started the first loop that will iterate through every epoch. Now you need to create the second loop, where the AI performs actions, updates the environment, and trains your CNN. Start it with these lines:

```
    # Starting the second loop in which we play the game and teach our
AI
    while not gameOver:

        # Choosing an action to play
        if np.random.rand() < epsilon:
            action = np.random.randint(0, 4)
        else:
            qvalues = model.predict(currentState)[0]
            action = np.argmax(qvalues)
```

As I mentioned, this is the loop in which your AI makes decisions, moves, and updates the environment. You start off by initializing a `while` loop that will be executed as long as you haven't lost; that is, as long as `gameOver` is set to `False`.

Then, you can see `if` conditions. This is where your AI will make decisions. If a random value from range (0,1) is lower than the epsilon, then a random action will be performed. Otherwise, you predict the Q-values based on the current state of the game and from these Q-values you take the index with the highest Q-value. This will be the action performed by your AI.

Then, you have to update your environment:

```
# Updating the environment
state, reward, gameOver = env.step(action)
```

You use the `step` method from your `Environment` class object. It takes one argument, which is the action that you perform. It also returns the new frame obtained from your game after performing this action along with the reward obtained and the game over information. You'll use these variables soon.

Keep in mind, that this method returns a single 2D frame from your game. This means that you have to add this new frame to your `nextState` and remove the last one. You do this with these lines:

```
# Adding new game frame to the next state and deleting the
oldest frame from next state
state = np.reshape(state, (1, env.nRows, env.nColumns, 1))
nextState = np.append(nextState, state, axis = 3)
nextState = np.delete(nextState, 0, axis = 3)
```

As you can see, first you reshape `state` because it is 2D, while both `currentState` and `nextState` are 4D. Then you add this new, reshaped frame to `nextState` along the 3rd axis. Why 3rd? That's because the 3rd index refers to the 4th dimension of this array, which keeps the 2D frames inside. In the last line you simply delete the first frame from `nextState`, which has index 0 (the oldest frames are kept on the lowest indexes).

Now, you can `remember` this transition in your experience replay memory, and train your model from a random batch of this memory. You do that with these lines:

```
# Remembering the transition and training our AI
dqn.remember([currentState, action, reward, nextState],
gameOver)
inputs, targets = dqn.get_batch(model, batchSize)
model.train_on_batch(inputs, targets)
```

In the first line, you append this transition to the memory. It contains information about the game state before taking the action (`currentState`), the action that was taken (`action`), the reward gained (`reward`), and the game state after taking this action (`nextState`). You also remember the `gameOver` status. In the following two lines, you take a random batch of inputs and targets from your memory, and train your model on them.

Having done that, you can check if your snake has collected an apple and update `currentState`. You can do that with these lines:

```
# Checking whether we have collected an apple and updating the
current state
```

```
        if env.collected:
            nCollected += 1


        currentState = nextState
```

In the first two lines, you check whether the snake has collected an apple and if it has, you increase nCollected. Then you update currentState by setting its values to the ones of nextState.

Now, you can quit this loop. You still have a couple of things to do:

```
    # Checking if a record of apples eaten in a around was beaten and
if yes then saving the model
    if nCollected > maxNCollected and nCollected > 2:
        maxNCollected = nCollected
        model.save(filepathToSave)


    totNCollected += nCollected
    nCollected = 0
```

You check if you've beaten the record for the number of apples eaten in a round (this number has to be bigger than 2) and if you did, you update the record and save your current model to the file path you specified before. You also increase totNCollected and reset nCollected to 0 for the next game.

Then, after 100 games, you show the average score, like this:

```
    # Showing the results each 100 games
    if epoch % 100 == 0 and epoch != 0:
        scores.append(totNCollected / 100)
        totNCollected = 0
        plt.plot(scores)
        plt.xlabel('Epoch / 100')
        plt.ylabel('Average Score')
        plt.savefig('stats.png')
        plt.close()
```

You have a list called scores, where you store the average score after 100 games. You append a new value to it and then reset this value. Then you show scores on a graph, using the Matplotlib library that you imported before. This graph is saved in stats.png every 100 games/epochs.

Then you lower the epsilon, like so:

```
    # Lowering the epsilon
    if epsilon > minEpsilon:
        epsilon -= epsilonDecayRate
```

With the `if` condition, you make sure that the epsilon doesn't go lower than the minimum threshold.

In the last line, you display some additional information about every single game, like this:

```
# Showing the results each game
print('Epoch: ' + str(epoch) + ' Current Best: ' +
str(maxNCollected) + ' Epsilon: {:.5f}'.format(epsilon))
```

You display the current epoch (game), the current record for the number of apples collected in one game, and the current epsilon.

That's it! Congratulations! You've just built a function that will train your model. Remember that this training goes on infinitely until you decide it's finished. When you're satisfied with it, you'll want to test it. For that, you need a short file to test your model. Let's do it!

Step 5 – Testing the AI

This will be a very short section, so don't worry. You'll be running this code in just a moment!

As always, you start by importing the libraries you need:

```
# Importing the libraries
from environment import Environment
from brain import Brain
import numpy as np
```

This time you won't be using the DQN memory nor the Matplotlib library, and therefore you don't import them.

You also need to specify some hyperparameters, like this:

```
# Defining the parameters
nLastStates = 4
filepathToOpen = 'model.h5'
slowdown = 75
```

You'll need `nLastStates` later in this code. You also created a file path to the model that you'll test. Finally, there's also a variable that you'll use to specify the wait time after every move, so that you can clearly see how your AI performs.

Once again, you create some useful objects, like an `Environment` and a `Brain`:

```
# Creating the Environment and the Brain
```

```
env = Environment(slowdown)
brain = Brain((env.nRows, env.nColumns, nLastStates))
model = brain.loadModel(filepathToOpen)
```

Into the brackets of the `Environment`, you input the `slowdown`, because that's the argument that this class takes. You also create an object of the `Brain` class, but this time, you don't specify the learning rate, since you won't be training your model. In the final line you load a pre-trained model using the `loadModel` method from the `Brain` class. This method takes one argument, which is the file path from which you load the model.

Once again, you need a function to reset states. You can use the same one as before, so just copy and paste these lines:

```
# Making a function that will reset game states
def resetStates():
    currentState = np.zeros((1, env.nRows, env.nColumns, nLastStates))

    for i in range(nLastStates):
        currentState[:,:,:,i] = env.screenMap

    return currentState, currentState
```

Now, you can enter the main `while` loop like before. This time, however, you won't define any variables, since you don't need any:

```
# Starting the main loop
while True:
    # Resetting the game and the game states
    env.reset()
    currentState, nextState = resetStates()
    gameOver = False
```

As you can see, you've started this infinite `while` loop. Once again, you have to restart the environment, the states, and the game over, every iteration.

Now, you can enter the game's `while` loop, where you take actions, update the environment, and so on:

```
    # Playing the game
    while not gameOver:

        # Choosing an action to play
        qvalues = model.predict(currentState)[0]
        action = np.argmax(qvalues)
```

This time, you don't need any `if` statements. After all, you're testing your AI, so you mustn't have any random actions here.

Once again, you update the environment:

```
# Updating the environment
state, _, gameOver = env.step(action)
```

You don't really care about the reward, so just place "_" instead of `reward`. The environment still returns the frame after taking an action, along with the information about game over.

Due to this fact, you need to reshape your `state` and update `nextState` in the same way as before:

```
# Adding new game frame to next state and deleting the oldest
one from next state
state = np.reshape(state, (1, env.nRows, env.nColumns, 1))
nextState = np.append(nextState, state, axis = 3)
nextState = np.delete(nextState, 0, axis = 3)
```

In the final line, you need to update `currentState` as you did in the other file:

```
# Updating current state
currentState = nextState
```

That's the end of coding for this section! This isn't, however, the end of this chapter. You still have to run the code.

The demo

Unfortunately, due to PyGame not being supported by Google Colab, you'll need to use Anaconda.

Thankfully, you should have it installed after *Chapter 10, AI for Autonomous Vehicles – Build a Self-Driving Car*, so it'll be easier to install the required packages and libraries.

Installation

First, create a new virtual environment inside Anaconda. This time, I'll walk you through the installation on the Anaconda Prompt from a PC, so that you can all see how it's done from any system.

Windows users, please open the Anaconda Prompt on your PC, and Mac/Linux users, please open your Terminal on Mac/Linux. Then type:

```
conda create -n snake python=3.6
```

Just like so:

Then, hit *Enter* on your keyboard. You should get something more or less like this:

```
(base) C:\Users\janwa>conda create -n snake python=3.6
Collecting package metadata (repodata.json): done
Solving environment: done

## Package Plan ##

  environment location: C:\Users\janwa\Anaconda3\envs\snake

  added / updated specs:
    - python=3.6


The following NEW packages will be INSTALLED:

  certifi            pkgs/main/win-64::certifi-2019.6.16-py36_0
  pip                pkgs/main/win-64::pip-19.1.1-py36_0
  python             pkgs/main/win-64::python-3.6.8-h9f7ef89_7
  setuptools         pkgs/main/win-64::setuptools-41.0.1-py36_0
  sqlite             pkgs/main/win-64::sqlite-3.29.0-he774522_0
  vc                 pkgs/main/win-64::vc-14.1-h0510ff6_4
  vs2015_runtime     pkgs/main/win-64::vs2015_runtime-14.15.26706-h3a45250_4
  wheel              pkgs/main/win-64::wheel-0.33.4-py36_0
  wincertstore       pkgs/main/win-64::wincertstore-0.2-py36h7fe50ca_0


Proceed ([y]/n)?
```

Type y on your keyboard and hit *Enter* once again. After everything gets installed, type this in your Anaconda Prompt:

```
conda activate snake
```

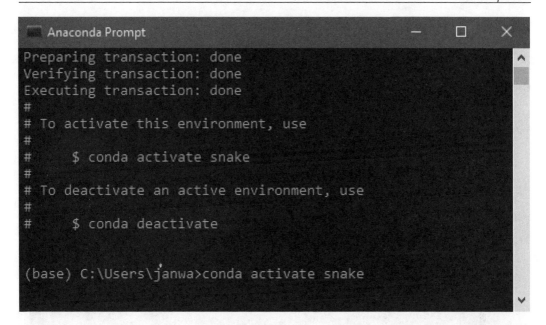

And hit *Enter* once again. Now on the left, you should see **snake** written instead of **base**. This means that you're in the newly created Anaconda environment.

Now you need to install the required libraries. The first one is Keras:

```
conda install -c conda-forge keras
```

```
Anaconda Prompt                                    —    □    X
Executing transaction: done
#
# To activate this environment, use
#
#     $ conda activate snake
#
# To deactivate an active environment, use
#
#     $ conda deactivate

(base) C:\Users\janwa>conda activate snake

(snake) C:\Users\janwa>conda install -c conda-forge keras_
```

After writing that, hit *Enter*. When you get this:

Type y once again and hit *Enter* once again. Once you have it installed, you need to install PyGame and Matplotlib.

The first one can be installed by entering `pip install pygame`, while the second one can be installed by entering `pip install matplotlib`. The installation follows the same procedure as you just took to install Keras.

Ok, now you can run your code!

If you've accidentally closed your Anaconda Prompt/Terminal for any reason, re-open it and type in this to activate the `snake` environment that we have just created:

```
conda activate snake
```

And then hit *Enter*. I got a bunch of warnings after doing this, and you may see similar warnings as well, but don't worry about them:

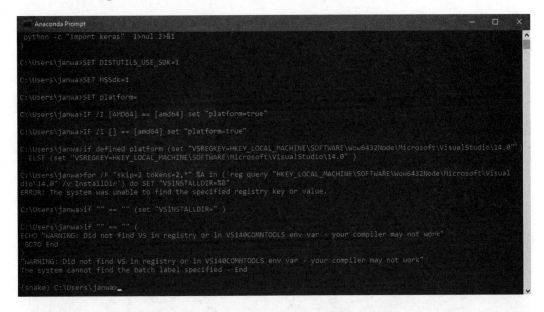

Now, you need to navigate this console to the folder that contains the file you want to run, in this case train.py. I recommend that you put all the code of Chapter 13 in one folder called Snake on your desktop. Then you'll be able to follow the exact instructions that I'll give you now. To navigate to this folder, you'll need to use cd commands.

First, navigate to the desktop by running cd Desktop, like this:

And then enter the `Snake` folder that you created. Just as with the previous command, run `cd Snake`, like this:

You're getting super close. To train a new model, you need to type:

```
python train.py
```

And hit *Enter*. This is more or less what you should see:

You have both a window on the left with the game, and one on the right with the terminal informing you about every game (every epoch).

Congratulations! You just smashed the code of this chapter and built an AI for Snake. Be patient with it though! Training it may take up a couple of hours.

So, what kind of results can you expect?

The results

Firstly, make sure to follow the results also on your Anaconda Prompt/Terminal, epoch by epoch. An epoch is one game played. After thousands of games (epochs), you'll see the score increase, as well as the snake size increase.

After thousands of epochs of training, while the snake doesn't fill in the entire map, your AI plays on a level comparable with humans. Here are some pictures after 25,000 epochs.

Figure 5: Results example 1

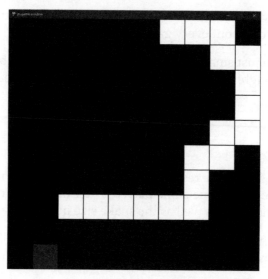

Figure 6: Results example 2

You'll also get a graph created in the folder (stats.png) showing the average score over the epochs. Here is the graph I got when training our AI over 25,000 epochs:

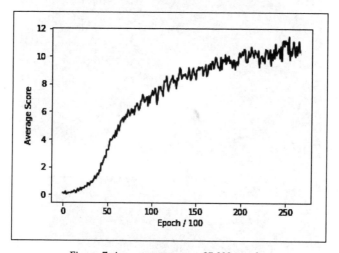

Figure 7: Average score over 25,000 epochs

You can see that our AI reached an average score of 10-11 per game. This isn't bad considering that before training it knew absolutely nothing about the game.

You can also see the same results if you run the `test.py` file using the pre-trained model `model.h5` attached to this chapter in GitHub. To do this, you simply need to enter in your Anaconda Prompt/Terminal (still in the same `Snake` folder on your desktop that contains all the code of `Chapter 13`, and still inside the `snake` virtual environment):

```
python test.py
```

If you want to test your model after training, you simply need to replace `model.h5` with `model2.h5` in the `test.py` file. That's because during the training the weights of your AI's neural network will be saved into a file named `model2.h5`. Then re-enter `python test.py` in your Anaconda Prompt/Terminal, and enjoy your own results.

Summary

In this last practical chapter of the book, we built a deep convolutional Q-Learning model for Snake. Before we built anything, we had to define what our AI would see. We established that we needed to stack multiple frames, so that our AI would see the continuity of its moves. This was the input to our Convolutional Neural Network. The outputs were the Q-values corresponding to each of the four possible moves: going up, going down, going left, and going right. We rewarded our AI for eating an apple, punished it for losing, and punished it slightly for performing any action (the living penalty). Having run 25,000 games, we can see that our AI is able to eat 10-11 apples per game.

I hope you enjoyed it!

14
Recap and Conclusion

In this final chapter, I'll provide you with a recap of the general AI framework for reference, and give you some words of advice as you take your work in AI to the next level. You've come a long way, and there's so much further you can take your AI studies in the future!

Recap – The general AI framework/ blueprint

Let's recap and provide the whole AI blueprint, so that you can refer to it whenever you need. You can even print it out and put it on your wall!

Step 1 – Building the environment

1. **Step 1-1**: Introducing and initializing all the parameters and variables of the environment.

2. **Step 1-2**: Making a method that updates the environment right after the AI plays an action.

3. **Step 1-3**: Making a method that resets the environment.

4. **Step 1-4**: Making a method that gives us at any time the current state, the last reward obtained, and whether the game is over.

Step 2 – Building the brain

1. **Step 2-1**: Building the input layer composed of the input states.

2. **Step 2-2**: Building the hidden layers with a chosen number of these layers and neurons inside each, fully connected to the input layer and between each other.

3. **Step 2-3**: Building the output layer, fully connected to the last hidden layer.

4. **Step 2-4**: Assembling the full architecture inside a model object.

5. **Step 2-5**: Compiling the model with a mean-squared error loss function and a chosen optimizer.

Step 3 – Implementing the Deep Reinforcement Learning algorithm

1. **Step 3-1**: Introducing and initializing all the parameters and variables of the Deep Q-Learning neural Network (DQN) model.

2. **Step 3-2**: Making a method that builds the memory in experience replay.

3. **Step 3-3**: Making a method that builds and returns two batches of inputs and targets, each one having `batch_size` elements.

Step 4 – Training the AI

1. **Step 4-1**: Building the environment by creating an object of the `Environment` class built in Step 1.

2. **Step 4-2**: Building the artificial brain by creating an object of the `Brain` class built in Step 2.

3. **Step 4-3**: Building the DQN model by creating an object of the `DQN` class built in Step 3.

4. **Step 4-4**: Choosing the training mode.

5. **Step 4-5**: Starting the training with a `for` loop over a chosen number of epochs.

6. **Step 4-6**: During each epoch, we repeat the whole deep Q-learning process, while also doing some exploration 30% of the time.

Step 5 – Testing the AI

1. **Step 5-1**: Building a new environment by creating an object of the `Environment` class built in Step 1.

2. **Step 5-2**: Loading the artificial brain with its pre-trained weights from the previous training.

3. **Step 5-3**: Choosing the inference mode.

4. **Step 5-4**: Starting the simulation.

5. **Step 5-5**: At each iteration (each minute), our AI only plays the action that results from its prediction, and no exploration or Deep Q-Learning training is happening whatsoever.

Exploring what's next for you in AI

You've come such a long way! Let's take a last step back and see what knowledge you've gained and what skills you've acquired:

- You have a solid intuition of Reinforcement Learning.
- You can use it to solve real-world problems.
- You can program in a way that sets you apart and puts you at the cutting edge of AI.
- You can write systems that learn and improve over time.
- You have the solid basics that allow you to go further in AI.

Speaking of going further, the question is: how? How will you apply what you've learned? What will you do next? First of all, your next step is:

Practice, practice, and practice

There are many ways to practice your AI skills. You can enter AI competitions like the ones on Kaggle, which contain problems that can be solved with deep reinforcement learning. You could build some new AIs like the one we created for the self-driving car. For example, you could build an AI with Kivy that plays the game of pong. There's a great AI platform, called **OpenAI Gym**, where you can practice building AIs for many types of applications, including:

- An AI that plays Atari games (Breakout, Pacman, Space Invaders, and so on).
- An AI that plays car racing.
- An AI that plays the game Doom.
- Training a virtual robot on how to walk and run.

I really recommend that you check out the Open AI Gym website, with all these fantastic applications you can work and practice on.

Going further, what would be the next step for you to take in making an impact in this world? Do you remember the 10 application fields of AI that we identified and explained in the introduction? Just pick your favorite! Pick the one that resonates the most with you. Let's remind you what they were:

1. Energy
2. Healthcare
3. Transport and logistics

4. Education

5. Security

6. Employment

7. Smart homes and robots

8. Entertainment and happiness

9. Environment

10. Economy and business

If you have an interest or passion in one of these fields, or even better some domain knowledge, you can combine that with your new AI skills to solve some problems in these industries. You can increase your impact by working with or in some tech companies, or by building your own one. There will always be massive demand for AI in each of these fields, which will always open many doors to you.

Speaking of open doors, that brings me to another next step I recommend to you.

Networking

Practicing is a necessity, but it's definitely not enough to make an impact with AI in this world. You also have to network. Whether it's working for a tech company, a company in another industry with an AI team, or for your own business, you should always network. This will open new doors, seed new opportunities, and increase your chances of success.

Networking today is easy. There are many AI events and conferences that you can visit, and the closest one will never be too far from your place. If you can't get to one, you can easily organize some AI meetups and after-work sessions yourself, where you discuss AI with other passionate people. You can also create groups using social media, where you can exchange ideas, brainstorm AI problems, and perhaps establish new connections through which you form synergy. Again, the more you network, the more you get all these benefits: connections, generated ideas, synergies, opportunities, opened doors, and AI journeys.

I want to finish this book by giving you my best advice for your professional life. The final recommendation I want to give you is the following.

Never stop learning

AI is a fast-evolving world so you must keep up to date on the latest state-of-the-art features. This book doesn't cover the latest breakthroughs in AI, but has given you the right basics and intuition for you to approach the latest developments in AI with confidence.

The right basics are a necessity, but not enough for you to keep up over time. What you must do is never stop learning. The good news is that it's easy to keep learning today! There are many great MOOCs that cover the latest advanced models in AI, as well as articles, research papers, blogs, not to mention YouTube videos where you can find the entirety of whole AI theory explained from scratch up to the state-of-the-art models. You have plenty of options to feed your brain with up-to-date AI knowledge. Just make sure you don't pick the worst reviewed content, and you'll be fine.

Let's recap. What are your next steps after this book?

1. Practice, practice, and practice.
2. Cross your AI skills with a field of application that resonates with you the most.
3. Network.
4. Never stop learning.
5. And, of course, keep up the hard work!

Yes, hard work will always be essential. Remember this. Success is only the tip of the iceberg, under which is hidden a tremendous amount of hard work. But don't worry; as soon as you feel passionate about your work and the purpose you follow, work will never be too hard. It will, in fact, feel effortless. That's why my recommendation number 2 here is very important: if you manage to pick a domain that resonates with your purpose, then you'll have found your way to make an impact with passion. If your passion is pure AI, even better! Then you can leverage it to solve problems and tackle challenges in several fields of applications, which gives you the amazing opportunity to have a diversified career.

On that note, I want to wish you a fantastic and very successful career. It was a great pleasure writing this book for you; my purpose is to democratize AI, and raise awareness among everyone that AI is an accessible technology that can make a difference for the better in this world. Thank you so much, and enjoy AI!

Other Books You May Enjoy

If you enjoyed this book, you may be interested in these other books by Packt:

Python Machine Learning - Third Edition

Sebastian Raschka, Vahid Mirjalili

ISBN: 978-1-78995-575-0

- Master the frameworks, models, and techniques that enable machines to 'learn' from data
- Use scikit-learn for machine learning and TensorFlow for deep learning
- Apply machine learning to image classification, sentiment analysis, intelligent web applications, and more

- Build and train neural networks, GANs, and other models
- Add machine intelligence to web applications
- Clean and prepare data for machine learning
- Classify images using deep convolutional neural networks
- Best practices for evaluating and tuning models
- Predict continuous target outcomes using regression analysis
- Uncover hidden patterns and structures in data with clustering
- Dig deeper into textual and social media data using sentiment analysis

Dancing with Qubits

Robert S. Sutor

ISBN: 978-1-83882-736-6

- See how Quantum Computing works, what makes it different, and why it could be so powerful
- Discover the complex, mind-bending mechanics that underpin quantum systems
- Understand the necessary concepts behind classical and quantum computing
- Refresh and extend your grasp of computing, quantum theory, and quantum computing
- Explore the main applications of quantum computing to scientific computing, AI, and elsewhere
- Comprehend the detailed overview of qubits, quantum circuits, and quantum algorithm

Leave a review - let other readers know what you think

Please share your thoughts on this book with others by leaving a review on the site that you bought it from. If you purchased the book from Amazon, please leave us an honest review on this book's Amazon page. This is vital so that other potential readers can see and use your unbiased opinion to make purchasing decisions, we can understand what our customers think about our products, and our authors can see your feedback on the title that they have worked with Packt to create. It will only take a few minutes of your time, but is valuable to other potential customers, our authors, and Packt. Thank you!

Index

A

activation function
about 128, 129
rectifier activation function 131-133
sigmoid activation function 130, 131
threshold activation function 129

AI
networking 330
skills, practicing 329, 330

AI blueprint
Deep Reinforcement Learning algorithm,
    implementing 328
environment, building 327
testing 328
training 328

AI, for cost implications
brain, building 269
environment, building 268
testing 270
training 269

AI models
about 2, 3
Deep Q-Learning 3
Q-Learning 3

AI, real-world business problem
about 220, 221
artificial neural network 221-223
brain, building 232, 233
deep reinforcement learning algorithm,
    implementing 238-245
demonstrating 258-268
environment, building 224-231
implementation 223, 224
implementing, with dropout regularization
    technique 237, 238

implementing, without dropout regularization
    technique 233-237
testing 256-258
training 245, 246
training, with early stopping 254, 255
training, without early stopping 246-253

AI solution
building, with Q-Learning 101-103
overview 66, 67, 168, 169

AI solution refresher
about 96
initialization (first iteration) 96
iterations 96, 97

AI solution, Snake game
about 293
brain 293, 294
experience replay memory 295

Anaconda
about 189
installation link 189
installing 189, 190

argmax method 147

arrays 20-22

artificial intelligence (AI)
about 1, 4
adding, value to business 6
companion robots 6
education 5
employment 5
energy consumption 4
environment 6
global economy 6
healthcare 4
journey 1, 2
models 2
robots 5

Q

Q-Learning
about 3, 91
AI fundamentals 77, 78
applying, to maze 78
implementing 97, 98
inference mode 103-105
used, for building AI Solution 101-103

Q-Learning, maze
actions 80, 81
AI, building 85
applying 78
Bellman equation 87, 88
environment, building 79
Q-value 85
reinforcement Intuition 88
rewards 81-84
states, defining 79
temporal difference 86, 87

Q-Learning process
about 88
inference mode 89
training mode 89

R

real-world business problem
environment, building 208
solving 207, 208
rectifier activation function 131
Reinforcement Learning
about 33, 34
principles 34
reference link 34
reward attribution
automating 105-108

S

self-driving car
Deep Q-Learning, implementing 177-188
demonstrating 189
environment, building 153
experience replay, implementing 175-177
implementing 169
Kivy, installing 194-204
libraries, importing 170

neural network architecture, creating 171-174
PyTorch, installing 192, 193
virtual environment, creating with Python 3. 190-192
server environment
actions, defining 214
assumptions 209
final simulation example 216-219
overall functioning 212-214
parameters 208
rewards, defining 215, 216
simulation 211
states, defining 214
variables 208
sigmoid activation function 130
simulation
environment, building 61-63
running 64-66
Snake game
about 288
AI, testing 315-317
AI, training 309-315
Anaconda Prompt, installing 318-323
brain, building 303-306
demonstrating 317
environment, building 288-303
experience replay memory, building 307, 308
implementing 295, 296
results 323-325
Softmax method 147
Standard model
versus Thompson Sampling model 57, 58
states 35
Stochastic Gradient Descent (SGD) 143-145

T

TensorFlow 169
text
displaying 18
displaying, with print() method 18
Thompson Sampling model
about 3, 42-48
actions, simulating 66
coding 43-73
distribution 48-52
implementation 68

V

W